Workview Office Student Edition

Schematic Entry and Digital Analysis

R. James Duckworth

Worcester Polytechnic Institute

PRENTICE HALL, Upper Saddle River, New Jersey 07458

Library of Congress Cataloging-in-Publication Data

Duckworth, R. James.
    Workview office student edition : schematic entry and digital
analysis / R. James Duckworth.
      p. cm.
    Includes bibliographical references and index.
    ISBN 0–13–490327–7
    1. Electronic circuit design—Data processing.   2. Computer aided
design.   3. Electronic apparatus and appliances—Design—Data
processing.   4. Workview office.   I. Title.
TK7867.D73   1997   96–42298
621.39′5′078—dc20   CIP

Acquisitions editor: **Tom Robbins**
Editorial/production supervision
  and interior design: **Sharyn Vitrano**
Editor-in-chief: **Marcia Horton**
Managing editor: **Bayani Mendoza DeLeon**
Copy editor: **Virginia Dunn**
Cover design: **ViewLogic Systems, Inc.**
Director of production and manufacturing: **David W. Riccardi**
Mnufacturing buyer: **Donna Sullivan**
Editorial assistant: **Nancy Garcia**

© 1997 by Prentice-Hall, Inc.
Simon & Schuster/A Viacom Company
Upper Saddle River, NJ 07458

Printed in the United States of America

10 9 8 7 6 5 4 3 2 1

ISBN 0-13-490327-7

PRENTICE-HALL INTERNATIONAL (UK) LIMITED, *London*
PRENTICE-HALL OF AUSTRALIA PTY. LIMITED, *Sydney*
PRENTICE-HALL CANADA INC., *Toronto*
PRENTICE-HALL HISPANOAMERICANA, S.A., *Mexico*
PRENTICE-HALL OF INDIA PRIVATE LIMITED, *New Delhi*
PRENTICE-HALL OF JAPAN, INC., *Tokyo*
SIMON & SCHUSTER ASIA PTE. LTD., *Singapore*
EDITORA PRENTICE-HALL DO BRASIL, LTDA., *Rio de Janeiro*

TRADEMARK INFORMATION

Viewlogic, when referring to products, is a
  registered trademark of Viewlogic Systems, Inc.
ViewDraw and *ViewSim* are registered trademarks of
  Viewlogic Systems, Inc.
ViewTrace and *Workview Office* are trademarks of
  Viewlogic Systems, Inc.
Windows, Windows NT, and Windows 3.1 are
  trademarks of Microsoft Corporation.
Speedwave is a trademark of Vantage.
VCS is a trademark of Chronologic Simulation.
MOTIVE and XTK are trademarks of Quad Design.

To Gail and Robert

Contents

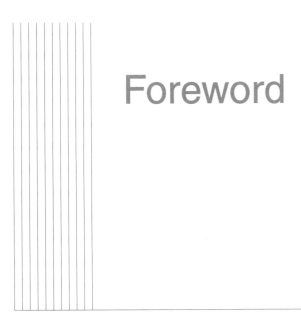

Foreword

Evolving technologies, increasing complexity, and time-to-market pressures are re-drawing the electronic design landscape. Viewlogic is dedicated to providing customers with the cutting-edge tools and solutions needed to push the design envelope as far as possible. But we've gone further to give our customers what they need by helping to develop the engineers to use the tools.

The education of tomorrow's design engineers is vital to our customers' success and to the future of the electronic design industry. These engineers are one of the most valuable assets in which our customers will invest.

Viewlogic is committed to the development of this resource through both our education program and our support of independent projects such as this book. We understand that the cost of education is rising faster than tuition and that the cost of leading-edge tools may be out of reach of a university's budget. The power and user-friendliness of Viewlogic's tools make them an ideal fit for the design world and the education environment.

Viewlogic is very pleased to be able to offer the *Workview Office*™ design environment for use in your digital design course. Experience with EDA tools is an important competitive advantage in today's job market. We're confident that this *Workview Office Student Edition* release will provide you with the kind of real-world design experience that you will be able to build upon throughout your career.

Alain Hanover
CEO of Viewlogic Systems, Inc

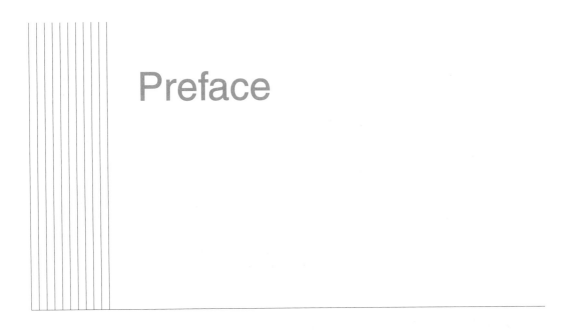

Preface

The text and software tools were developed for use in electrical engineering, computer engineering, and computer science courses dealing with the design and analysis of digital circuits. This book is a unique combination of logic design exercises and powerful computer-aided engineering software from Viewlogic Systems, Inc. The class-tested laboratory exercises provide realistic examples of combinational and sequential circuits and are used in a tutorial manner to demonstrate how to use the design entry and design analysis tools.

Instead of students having to design circuits by hand and build them in a conventional lab to see if they work, students now have access to the same type of design tools used by engineers. The software tools provided with this book are not the artificial and limited programs usually distributed in student versions. The versions of the Viewlogic® *ViewDraw*®, *ViewSim*® and *ViewTrace*™ tools provided are identical to the commercial tools sold by Viewlogic for many thousands of dollars; the only restriction is that they are limited in the size of circuits supported. Other student CAD tools are artificial and have major limitations. Most do not consider real device delays or simply ignore timing altogether, while others assume that every device has a 'unit' delay. In either case, the results are meaningless. Without detailed timing simulation it is impossible to test a circuit correctly and have confidence that it would work well if built.

Students can draw their circuit design using the *ViewDraw* schematic editor and then perform extensive analysis by using the *ViewSim* gate level simulator. As the circuit is tested by applying various input combinations using *ViewSim*, the circuit drawing is back-annotated with the new digital logic values at all the nodes of the circuit. This provides students with good feedback on how their circuit is actually working and makes faultfinding easy. If bugs are found it is a simple matter to correct the schematic. The waveform analyzer tool, *ViewTrace*, can also be used to provide a graphical display of the signals changing with time.

Once the circuit is performing as required, the students could still build the circuit, or a portion of the circuit, in a regular lab. The major difference is that the circuit should now work the first time!

Using commercial quality tools for digital logic design as provided by *Workview Office Student Edition* allows students to spend their time on understanding the theory and design issues rather than spending long hours manually constructing and debugging circuits in the lab.

Although this book was primarily written for students using the student edition release of *Workview Office*, most of the material, including the tutorials, should be useful for any engineer using the commercial version of *Workview Office*.

Organization of the Book

This book is divided into three parts. Part One provides a complete overview of the rapidly growing and changing field of computer-aided engineering. A historical perspective is provided describing how digital systems were designed and analyzed before CAE tools were available. Then, the process of entering a design and simulating it is explained, while at the same time concepts such as back-annotation are introduced. The importance of rapid prototyping to shorten the product development stage is described. New developments and technologies such as programmable logic devices, hardware description languages, and logic synthesis are also covered in this section.

Part Two provides an introduction to the Viewlogic schematic entry and digital analysis tools provided with this text book, followed by an installation and a "quickstart" section if you just cannot wait to start using the software! The main section of Part Two consists of two comprehensive but simple tutorials to show students how to design and analyze their circuits using *Workview Office*. The tutorials are based on one of the laboratory exercises so students, as well as learning how to use the software tools, are also designing and analyzing a reasonably complex circuit. However, the tutorials have been written so that they could be used with a different design if that is required.

Also in Part Two is a chapter describing how to access and use the comprehensive on-line help, tutorials, and documentation for the *Workview Office* products.

Part Three consists of ten digital logic design laboratory exercises, commencing with simple combinational circuits and progressing to the design of finite state machines. In each exercise the student is provided with a "theoretical background" section to give an overview of the theory and design issues that will be incorporated in the laboratory exercises.

After a preliminary laboratory exercise to introduce students to logic devices and their operation, the second exercise involves the design of a four-bit binary ripple-carry adder to provide an introduction to the design of simple combinational circuits. This adder is used as the design exercise in the tutorials in Part Two to introduce the students to the design entry and analysis software tools provided with this textbook.

The remaining laboratory exercises develop more complex arithmetic and logic circuits. Once the ripple-carry adder has been completed, the students, through timing analysis of their circuits, should appreciate that this type of design (using ripple carry) is far too slow for anything but adding or subtracting a few digits. The design of a functionally equivalent circuit using look-ahead carry is then carried out. Again, though simulation and timing analysis, the students can now explore the improved performance capability of this new design.

In a subsequent laboratory exercise, the simple arithmetic unit is enhanced to perform logic functions such as logical AND and OR operations. At this point the students will have designed a reasonably complicated combinational circuit that is representative of arithmetic logic units (ALUs) found in many microprocessors.

Later laboratory exercises involve the design of sequential logic circuits. A simple pseudo-random number generator circuit in the form of a synchronous binary counter is used to introduce students to the design of finite state machines. Two laboratory exercises involve the design of traffic controllers, the second one having a car detector and a left turn indicator capability. Another laboratory exercise involves the design of a sequential combination lock, and the final design is of a vending machine.

In each laboratory exercise, once the design has been entered, the students verify that it is functioning correctly through simulation. An optional section requires the students to construct part of their design in a conventional laboratory setting, which allows them to complete the whole cycle from conceptual design to a physical implementation.

Preface to the Instructor

This book contains a tutorial introduction to the *Workview Office ViewDraw*, *ViewSim*, and *ViewTrace* tools. This student edition release of the tools provides the same functionality as the normal commercial version apart from a few restrictions on the size and type of designs that can be created and saved. The maximum number of modules that can be used in a design is limited to 300 and the names of the design files are limited to a set of 20. Ten of these design names were selected to support the laboratory exercises in this book (such as fourbit, traffic, vending), while 10 other names have been provided (design0 to design9) to support any other work.

The *Workview Office Student Edition* release contains *Builtin* and 74 series libraries. The *Builtin* and *74LS* libraries are used in the tutorials and are installed automatically. The other 74 series libraries are available on the CD-ROM and can be used to suppport your particular design exercises.

Acknowledgments

Many people at WPI, Viewlogic, and Prentice Hall have supported and helped me develop this text book and the associated software tools over the last few years. In particular, John Orr, my department head, has provided me with encouragement and given me the flexibility in my duties needed to complete such a major project.

The book and software would have been impossible to create if not for the support of Viewlogic Systems, Inc. This support started in 1993 when Viewlogic donated approximately $5 million in CAE tools for our PC and workstations, which we put to immediate use in our undergraduate and graduate classes. In addition, Alain Hanover, president and CEO, was very encouraging when I first discussed this textbook and student version of his tools. After initial discussions with Alain, my main contact at Viewlogic has been Preetinder Virk, who helped to provide the resources and effort needed to produce the student edition of the software.

Many other people at Viewlogic have helped me make this book and software into a successful product. I would like to thank Jim Meikle, the manager of educational services, for his support and enthusiasm to help get the project started. In the same department, Gus Eifrig has always been very helpful. In addition, Karen Wills, Public Relations, provided some material for the book.

I would also like to thank the anonymous reviewers of earlier versions of this book and software who provided many comments and suggestions.

And last but not least, I would like to thank David Ostrow, Sharyn Vitrano, and Tom Robbins at Prentice Hall. David is the sales representative for my school and has been one of the most helpful reps I have known. Sharyn Vitrano, production editor for this book, deserves a special mention for suffering through many last-minute revisions and changes as the student version of the software was being finalized. I would like to thank Tom, the executive editor at Prentice Hall, for his efforts and support in getting this project completed.

I hope you enjoy reading this text and using the Viewlogic tools and get as much fun and experience out of the laboratory exercises as I have had in using the software and writing this book.

R. James Duckworth
Worcester Polytechnic Institute

Workview Office
Student Edition

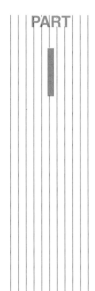

PART

I

INTRODUCTION

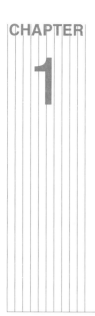

Introduction to Computer-Aided Engineering of Digital Circuits

This chapter provides an introduction to the use of computer-aided design (CAD) and computer-aided engineering (CAE) tools for the design and analysis of digital systems. It starts with a historical perspective showing the traditional design process and then explains how the CAE tools available today can be used to improve all aspects of the design process.

The design and entry of digital logic designs using schematic capture is explained, followed by a section describing the importance of carrying out both timing and functional analysis. The process of back-annotation of simulation values to the original schematic is also demonstrated.

Other sections explain the concept of hierarchical design, the advantages and necessity of rapid prototyping, the versatility of programmable logic devices and field programmable gate arrays (FPGAs), and the relatively new topics of hardware description languages and design synthesis.

Introduction

The use of computers to aid in the design and analysis of digital systems is relatively recent and is still undergoing major improvements and changes. To help explain this topic and to introduce the computer-aided engineering tools provided with this textbook I would like to provide some personal historical perspective.

Historical Perspective—The Traditional Design Process

About 10 years ago, when I was a design engineer for a company in England, I designed and built a variety of complex digital logic and microprocessor systems. One of the systems I designed and constructed could be used to test the quality and performance of video display monitors. The system had to produce various test patterns consisting of different characters at different resolutions and colors. This design required about 30 SSI and MSI (small and medium scale integration) devices, such as simple gates, flip-flops and shift registers. The bit maps of the characters that had to be displayed were stored in memory devices (EPROMS).

Then, there were very few tools available (computer or otherwise) to help in the design of digital and computer systems. The only tools I had were a pencil and paper and drawing stencil and a calculator. I started with a list of requirements for the system, from which I produced a block diagram (drawn on paper) identifying the main components. I then proceeded to take each of the blocks and selected SSI and MSI devices from my TTL data book to implement the required functionality. The next step involved drawing circuit diagrams (again by hand, using my logic gate stencil and ruler!) of each of the functional blocks, from which I finally obtained a circuit diagram of the complete system.

During this stage I referred continuously to data books to check the logical operation and timing characteristics of the components I was using to confirm that the circuit would work correctly. I first carried out a low-level functional analysis where I checked that the correct logic signals were generated and passed correctly from one component to another. My approach was to place logic levels (0's and 1's) at various inputs or intermediate nodes in my circuit and then work out what the logic levels would be at other places in the circuit to confirm that it was working correctly. This was a tedious and time-consuming process, but it was nevertheless an essential chore in an attempt to guarantee that the circuit would function correctly. If I found an error in my design then I had to redraw the schematic and repeat the process.

In addition to a functional analysis of the design, I had to carry out detailed timing analysis of the circuit to ensure that I was not violating any timing constraints on the integrated circuits and that the required signals were being passed from one place in the circuit to the next at the appropriate time. This was also a time-consuming exercise that involved looking up switching characteristics of all the components from the TTL and other data books, drawing timing diagrams (usually on graph paper) to ensure that all timing constraints were met. And this was not a particularly difficult or complex design—at least not by today's standards!

As mentioned above in my example design, there are a number of steps involved in the development of any digital system. These steps are shown in Figure 1-1.

Figure 1-1: Development of a digital system

As the name indicates, the design entry stage is where a description of the required system is entered. The entry process may involve graphical and/or textual input. The design verification stage is used to confirm that the completed design will operate correctly and usually involves extensive functional logic and timing simulation. The final stage is where the design is transferred onto the parts (the integrated circuits) that will be used to actually implement the design. Depending on the technology used for the implementation (standard parts, programmable logic, gate array, etc.), there may be a requirement for timing parameters to be passed back from the implementation stage to the verification stage. This path is shown as a dashed line in Figure 1-1 and is due to the fact that some of the timing information (routing delays, for example) is available only once the circuit has been placed on the integrated circuit(s) that will be used to implement the design.

Having completed the above overview, we can now investigate the use of computer-aided engineering tools in the system development process.

Computer-Aided Engineering

If I were to design a circuit such as the video test generator today or to design arithmetic circuits or traffic controllers like the exercises in this book, how would I go about it? Although the general procedure is still the same, we can now use computers and software programs to make our life easier. These programs are produced by many different companies and have a variety of different names and use many different terms and acronyms. In this section we review some of the programs from Viewlogic that allow us to design and analyze digital circuits.

Schematic Capture Using *ViewDraw*

Instead of drawing circuits by hand with paper, pencil, and stencils, we can use *schematic capture* drawing packages. With the aid of menus, keyboard commands, and the mouse it is now easy to draw circuit diagrams by placing component outlines from comprehensive libraries and connecting them with lines representing the wires. If you then want to modify your design, instead of having to start with a new piece of paper, it is usually a simple matter to make modifications to the electronic version. An example of a circuit diagram, also called a schematic, is the full adder shown in Figure 1-2. This was drawn using the Viewlogic schematic entry tool called *ViewDraw*, which is included with this textbook.

Components can be moved by simply clicking on the part and copying it, deleting it, or dragging it to a new place. When components are moved from one location to another the connecting wires can also be dragged at the same time. Most of the newer schematic entry packages such as *ViewDraw* take care of automatically routing the wires around other components while still retaining the required connectivity.

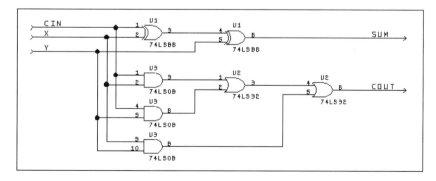

Figure 1-2: *Schematic of a simple circuit—a full adder*

Other useful features include the ability to zoom in and out of the diagram as you work on a specific section of the design, the automatic numbering of integrated circuits and other components (resistors, capacitors, etc.) as they are selected and placed, and the addition of text and other graphical items to annotate and describe the operation of the circuit.

The drawing process usually involves the following steps:

- Selecting the required components from libraries.
- Placing the components in the desired locations on the drawing.
- Connecting the components with lines representing the electrical connections.
- Labeling inputs and outputs.
- Adding text for the title of the drawing, including date, authorship, and any other descriptive information.

This section has described the design entry process using schematic capture. In a later section we show how hardware description languages (HDLs) may also be used to describe and enter a design description. We now want to investigate the tools and concepts involved in the design analysis and verification stage.

Design Analysis and Verification

Once a circuit description has been drawn using a schematic entry package, or entered in some other manner, it can be analyzed for correct operation. This requires a functional as well as a timing analysis. The functional analysis is usually static in nature. A set of input values are applied to the circuit and the resulting outputs are checked to confirm that the circuit responds appropriately. Before simulation tools were available, logic levels were applied by hand and then worked out by inspection for each of the intermediate and output nodes. The following kind of thought process would be involved: "If U3 pins 4 and 5 are logic '1' then pin 6 will also be a logic '1', which means that U2 pin 2 is also a logic '1' and so U2 pin 3 must be ....." For anything but the simplest circuit the novelty of this kind of analysis soon wears off! However, it was essential if one was to have any kind of confidence that the circuit would perform as expected once built.

As well as the obvious time-consuming and error-prone problems with this manual testing of the circuit, a more subtle problem arose because you had certain expectations in the way the circuit was supposed to work. It was quite easy to overlook some problems because you never considered the particular input combination that would cause the error. This could result in a circuit that did not work correctly once it was built because you had not thought that a certain logical combination or timing loop could exist.

You will be pleased to know that many simulation tools are also now available to automate the testing of your circuits. When a schematic is completed, a special file called a netlist is generated that contains a complete description of the circuit in a form that can be read by other computer programs, such as simulators or printed circuit board layout programs.

Functional and Timing Analysis Using *ViewSim* and *ViewTrace*

Many digital logic simulation programs are available, and they differ enormously in their capabilities and their ease of use. The better programs, for example, allow the application of input values in a variety of ways, from keyboard or file entry of input values to the use of timing waveforms. Once the inputs have been applied the simulator can then automatically work out the new logic values that will exist at every junction (or node) in the circuit. These values may be displayed in a listing form or now it is common for simulators to display them on the actual circuit diagram in a process known as *back-annotation*.

Figure 1-3 shows the previous full-adder circuit after a simulation cycle has been carried out. The input values applied to the circuit are shown in the square boxes on the left and the calculated values of the rest of the nodes are displayed as '1' and '0' values. It can be seen that the full adder produces a sum output of '0' and a carry-out of '1' when the carry-in is '0' and the two input bits, X and Y, are both '1'—which is what one would expect from such a circuit.

Using such a simulation tool, the process of functional testing is now much easier. Any errors in the expected output values can be quickly traced back to the part of the circuit causing the error and the circuit diagram can be quickly modified for retesting.

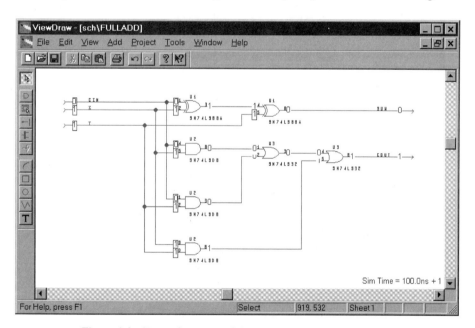

Figure 1-3: Circuit diagram with back-annotated simulation values

If you refer to Figure 1-3 you can see the current simulation time displayed in the bottom right corner, which for this example was 100 ns (The +1 refers to something called delta time, which is required for VHDL simulation—we can ignore this for our circuit simulation.) The *Workview Office* gate level simulator, *ViewSim*, which is one of the software tools included with this book, provides for full timing analysis as well as functional analysis. This means that the simulator calculates not only the correct logical values at all nodes in the circuit but also at what time the logic signals change values. The simulator does this by referring to the switching characteristics of all the devices that have been used in the circuit. Such an analysis is important because a purely static analysis of a circuit (where timing and propagation delays of signals are ignored) may indicate it will work correctly. However, the system could fail when constructed since real devices have important timing parameters that must be satisfied in order to function correctly.

The *ViewSim* simulator can provide the exact times at which all the nodes in a circuit change value. It can also detect timing violations on components. For example, consider a D-type flip-flop in which the Q output changes to reflect the value present on the D input after the clock changes from a low to a high level. We can consider two timing parameters: How long does it take from the positive clock transition to the Q output changing value and how long must the D input have been stable (i.e., not changing) before the positive clock transition? These timing parameters are shown as T_{PLH} and T_{SETUP}, respectively, in the timing diagram below:

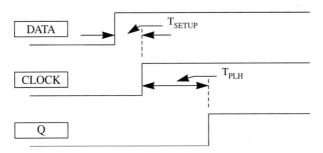

The actual timing of these two characteristics will depend on the type of device used for the flip-flop. For example, a part called 74LS74A, which has two D-type flip-flops, would have the following specifications:

	min	typ	max
DATA setup time before CLOCK (T_{SETUP})	20 ns		
Q output high after CLOCK (T_{PLH})		13 ns	25 ns
Q output low after CLOCK (T_{PHL})		25 ns	40 ns

whereas a 74S74 is functionally equivalent but made from a different technology and operates significantly faster, with these specifications:

	min	typ	max
DATA setup time before CLOCK (T_{SETUP})	3 ns		
Q output high after CLOCK (T_{PLH})		6 ns	9 ns
Q output low after CLOCK (T_{PHL})		6 ns	9 ns

The *Workview Office* simulation libraries contain these and other timing characteristics for all the devices it supports and during the simulation of your circuit it will check that none of the specifications are violated. For example, if you were using a 74LS74A and the D input was modified less than 20 ns before the clock transition, a warning message will be generated alerting you to the timing violation.

We can show the timing delays that occur during the switching of a circuit in the form of timing diagrams by using the waveform analyzer tool called *ViewTrace*. Referring back to the full adder circuit we analyzed earlier, we can obtain the timing diagram shown in Figure 1-4.

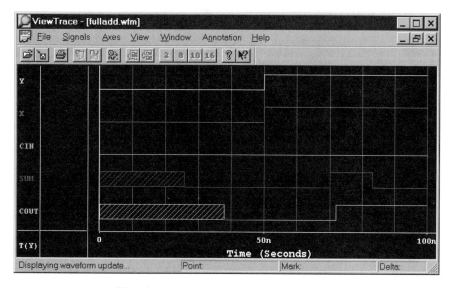

Figure 1-4: Timing diagram of full adder circuit

This diagram shows that the X and Y inputs were changed from low to high levels at 50 ns and the COUT changed from a low to a high level 22 ns later (at 72 ns). Also note that the SUM output changed temporarily from a low to a high level (at 70 ns) before stabilizing at a low level. The pulse generated (also known as a glitch or timing hazard) is a result of different delays through the circuit as the gates change values. A purely static analysis of this circuit would have shown only that the COUT changed to a logic '1' and the SUM output stayed at a logic '0'. If we have a larger circuit, then, depending on where the SUM output was connected, the pulse may or may not cause a problem downstream. With the timing analysis carried out by *ViewSim* we are aware that a potential timing problem exists and can take steps to correct it if necessary.

Also notice the shaded portions at the start of the SUM and COUT signal waveforms. These regions represent an unknown logic level and indicate that the simulator could not resolve what the actual logic level should be for these outputs during the time shown. This is because any inputs applied at the start of the simulation run will take time to propagate through the gates in the circuit before they produce a stable output. These unknown states are usually present at the start of a simulation cycle in any circuit and will last as long as it takes for the circuit to settle down to a stable state.

In this book you will use the three *Workview Office* tools—*ViewDraw*, *ViewSim*, and *ViewTrace*—to draw, simulate, and analyze a variety of digital logic circuits. Once the circuits are correctly analyzed, you should have a high degree of confidence that, if built, they would function correctly. In the following sections we review some newer developments in CAE and also explain some other terms and concepts.

Hierarchical Design

There are two different ways to design a circuit: One way is called a *flat* design and the other is called a *hierarchical* design. As an example, consider a microprocessor consisting of the following sections: an arithmetic logic unit (ALU), a microprogram store and sequencer, a register bank, and an instruction decode unit. One could design this microprocessor by starting with a (very) large sheet of paper on which all the gates and interconnections would be placed. Although this is theoretically possible, it would be very difficult to work with. There would be no easy way to divide up the design so that several engineers could work together on the overall design. Also, identifying the different blocks and their functionality would be nearly impossible. This type of *flat* design is therefore used only for small circuits.

The more usual method to carry out the design of anything but the most trivial circuit is to use a *hierarchical* approach. First, a block diagram is produced identifying the key functional elements required. Then each of these functional blocks is gradually refined in detail until the whole design is implemented. Figure 1-5 shows part of one possible hierarchical design of the microprocessor mentioned above.

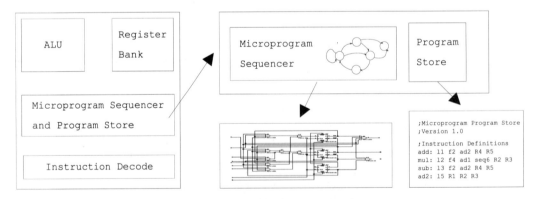

Figure 1-5: *Hierarchical view of an example microprocessor*

With such a design approach (also known as top-down design) engineers can work together on different sections of the design at the same time. One group, for example, may be responsible for the design of the ALU, while another group may be working on the design of the microprogram. This approach also provides for the capability of module reuse.

In a similar way that subroutines or functions developed for one program may be copied to another program, parts of the microprocessor circuit may be reused in another design.

A hierarchical approach also simplifies design verification. Instead of the engineers having to wait until the whole design is completed before testing can be carried out, each of the separate functional blocks may be individually analyzed and debugged as they are designed.

Rapid Prototyping

As recently as a few years ago, the time taken for the design and testing of a new product was relatively leisurely. Even if the time from concept to finished product was a few years, this pace was usually acceptable since the expected life-time of the product was relatively long. This is no longer true. With increased competition, new products must be continuously introduced with additional and enhanced features to enable one manufacturer to gain or maintain a competitive edge. This means that the time available to design, build, and test new products has been shortened from a year or more to just a few months.

Design engineers are thus looking for development tools and implementation technologies that enable them to meet these shorter deadlines. One of these is known as rapid prototyping. This means that once a product specification is completed, design engineers must produce a working prototype as quickly as possible. And, since the design specifications may continue to change, engineers may have to be able to modify their designs quickly and easily.

To achieve these goals, three main developments have made the job less difficult for the engineers:

- Improved computer-aided engineering and engineering design automation tools

- New types of user programmable logic devices

- The introduction of hardware description languages

The first development is the basis on which this book and software tools were produced. Computer-aided engineering tools are now used routinely by design engineers, and Viewlogic and I wanted to be able to provide students with access to the same type of professional quality software used in industry. Programmable logic and hardware description languages are described in the following sections.

Programmable Logic Devices

Over the last few years there has been tremendous progress made in the development of *user-programmable* logic devices. In conventional logic devices, the functionality is set by the manufacturer and cannot be changed. For example, a 74LS00 integrated circuit consists of four two-input NAND gates, while a 74LS11 contains three three-input AND gates. It is necessary to use these *standard parts* in combination to make a circuit with the required functionality. Programmable logic devices are more versatile, have a range of possible circuit functions, and may be customized by the design engineer (i.e., you!).

Earlier programmable logic devices (PLDs) consisted of an array of AND and OR gates in the form of a matrix connected by *fuses* that could be left intact or blown to implement various sum-of-product expressions. These devices were programmed using a *logic programmer* and a single PLD would be able to replace a few conventional SSI integrated circuits. In recent years many more PLDs have been produced with much larger capacities and with different internal architectures. Instead of the few tens of gates that the earlier PLDs contained, newer devices contain tens of thousands of gates and it is not uncommon for a complete logic circuit to just consist of one or two PLDs. As well as reducing the part count, another advantage of PLDs is the relative ease with which changes can be made to the circuit just by reprogramming the part—this is useful to correct bugs or to add new features to the product.

Manufacturers call their PLDs by different names to try and differentiate their products from those of their competitors. For example, the term EPLD is used for Erasable or EPROM-based Programmable Logic Device, FPGA for Field Programmable Gate Array, and CPLD for Complex Programmable Logic Device.

As with other logic technologies, the basic methodology for designing with programmable logic devices consists of a number of interrelated steps: design capture, simulation, synthesis, place and route, and verification as shown in Figure 1-6. The design capture step specifies the logic functions needed for a particular application. Examples of design capture methods include schematic entry (described earlier), truth tables, state diagrams, or text based descriptions.

Figure 1-6: Basic PLD design methodology

Once the design has been entered a functional simulation can usually be carried out to detect any major errors and the design description modified if necessary. The synthesis step converts the design entry description into a simplified set of logic equations which can be used to determine the logic resources required in the chosen PLD device. The next step is to determine an optimal placement and interconnect routing for those resources. This will have to take into account the internal architecture of the device and the location and quantity of individual gates, flip-flops, or other logic building blocks. The final verification step can include in-system testing, simulation, and timing analysis. Full timing simulation cannot occur until after place and route, since the actual timing characteristics of the design are dependent on how the design is implemented.

The design process invariably is iterative, returning to the design entry phase for correction and optimization. Typically, generic tools such as *ViewDraw*, *ViewSim* and *ViewTrace* are used for entry, simulation, and verification, but architecture-specific tools provided by the vendor of the programmable logic device are needed for the synthesis and place and route steps.

Hardware Description Languages

A recent development in the design of logic systems is the use of hardware description languages (HDLs) to describe the required behavior of a digital system. System designers can now describe at a high level of abstraction how they want the system to function. So-called synthesis tools can then convert this information into the necessary logic and circuit implementation. By letting the synthesis tools design the circuit, engineers no longer have to worry about which gates, shift registers, or flip-flops are needed for the implementation and can concentrate instead on the high-level functionality required by the product.

The use and application of HDLs and synthesis tools is similar to the use of high-level languages and compilers for developing a program. The majority of programmers now write their programs in high-level languages (such as C and Pascal), and they let the compiler produce the low-level machine instructions necessary to implement their algorithms. In fact, most programmers are probably not even aware of the types of machine instructions produced by the compiler. In a similar process, the goal of designing a logic system using a hardware description language is to let the synthesis tools (the equivalent of the compiler and assembler) worry about the low-level logic circuits required to implement the design. However, unlike the majority of compilers for high-level languages, which almost universally produce fast, optimized code, synthesis tools are still relatively new and the inexperienced user can easily produce inefficient or incorrect designs if not careful.

The two most common HDLs in current use are Verilog and VHDL. Verilog started out as a proprietary product developed by Cadence, but has since been declared a public domain language and is being submitted as an IEEE standard. VHDL was developed as an offshoot of the U.S. Government's VHSIC program as a standard language used to describe digital hardware devices, systems, and components. It was approved as an IEEE standard in 1987. VHDL is an acronym that stands for VHSIC Hardware Description Language; VHSIC stands for Very High Speed Integrated Circuit. VHDL is considered by many to be more verbose and difficult to use[1] than Verilog, but it is more flexible and expressive.

Figure 1-7 shows an example how VHDL may be used to model a digital component, in this case a two-to-one data selector or multiplexer.

The ENTITY section defines the external view of the component. The PORTS define the inputs and outputs of the entity and are like IC pins. The ARCHITECTURE defines the behavior or structure of the ENTITY. The VHDL code essentially says that if the *sel* input is a logic '0', then make the *y* output equal to the value on the *a* input; if the *sel* input is a logic '1', then make the *y* output equal to the value on the *b* input.

---

[1] Some people refer to VHDL as an acronym for Very Hard Description Language!

```
ENTITY mux is
  PORT (a, b, sel : IN bit;
        y : OUT bit);
END mux;

ARCHITECTURE behavior OF mux IS
BEGIN
  PROCESS(a, b, sel)
  BEGIN
    IF sel = '0' THEN
      y <= a;        -- read as "gets the value of"
    ELSE
      y <= b;
    END IF;
  END PROCESS;
END behavior;
```

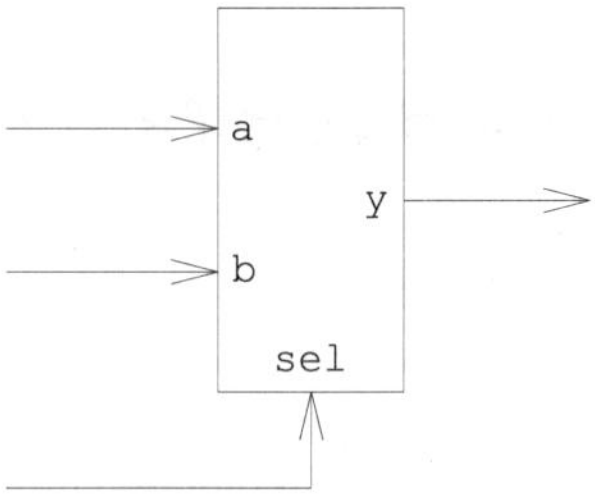

Figure 1-7: VHDL code and logic symbol for a 2-line to 1-line multiplexer

Some of the advantages of using HDLs for the design of systems are

- Documentation
- Flexibility (easier to make design changes or mods)
- Portability (if HDL is a standard)
- One language for modeling, simulation, and synthesis
- The synthesis tools worry about gate generation
- Productivity

We look at the process of logic synthesis in the next section. Although VHDL and other languages have great potential, the biggest drawback is the fact that they constitute a completely different way of approaching design. Many engineers are used to thinking and designing in graphical ways (block diagrams, flowcharts, and schematics) instead of in text.

Logic Synthesis

Synthesis is the process of obtaining a technology-dependent implementation from a circuit description. The phrase "technology-dependent" means that the implementation is targeted at a specific architecture. For example, the design of a traffic light controller could be implemented using standard off-the-shelf components such as TTL integrated circuits (flip-flops, AND, OR, or NAND gates) or it could be implemented in a programmable logic device or fabricated as a gate array. In each case the synthesis tools would have to map and partition the design into the architectural capabilities of the target technology. A PLD usually has a programmable AND-OR structure with configurable registered output logic blocks. Alternatively, a gate array usually consists of thousands of one type of simple gate (two-input NAND gates for example). In each case the synthesis tools must map the design to the available architecture so the circuit will function correctly whatever the final implementation.

Figure 1-8 shows the logic synthesis process, which takes a digital circuit description and translates it into a gate level design, optimized for a particular implementation technology.

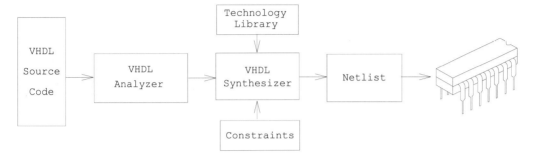

Figure 1-8: *Block diagram showing the process of logic synthesis*

As an example of this process we investigate how a digital circuit can be automatically produced (synthesized) from a VHDL description.

A logic symbol of a 2-line to 4-line decoder is shown in Figure 1-9. A truth table for the operation of this device is also shown in the figure. It can be seen that when the enable signal is high, the four output signals (Y0-Y3) will also be high. When the enable signal is low, one of the Y outputs will be low corresponding to the output selected by the combination of the two select lines.

INPUTS			OUTPUTS			
EN	SEL1	SEL0	Y3	Y2	Y1	Y0
H	X	X	H	H	H	H
L	L	L	H	H	H	L
L	L	H	H	H	L	H
L	H	L	H	L	H	H
L	H	H	L	H	H	H

Figure 1-9: *Logic symbol and truth table of a 2-line to 4-line decoder*

We can now write a VHDL description of the operation of this decoder. In conventional programming there are a number of different ways to code algorithms. The same is true when writing a VHDL description and there are different styles and techniques to writing VHDL models of components. An example VHDL program describing the behavior of the decoder is shown in Figure 1-10. (Note that comments in VHDL are preceded by two dashes --.)

```
-- Example 2-line to 4-line decoder (74139)
ENTITY decoder IS
  PORT(sel    : IN bit_vector(1 TO 0);     -- defines a 2-bit bus
       en     : IN bit;                    -- a single bit input
       y      : OUT bit_vector(3 to 0));   -- defines a 4-bit bus
END decoder;

ARCHITECTURE behavior OF decoder IS
BEGIN
  PROCESS(sel, en)
  BEGIN
    IF en = '0' THEN         -- if enable is low then
      CASE sel IS
        WHEN "00" =>
          y <= "1110";       -- y0 = low
        WHEN "01" =>
          y <= "1101";       -- y1 = low
        WHEN "10" =>
          y <= "1011";       -- y2 = low
        WHEN "11" =>
          y <= '0111";       -- y3 = low
      END CASE;
    ELSE
      y <= "1111";
    END IF;
  END PROCESS;
END behavior;
```

Figure 1-10: VHDL code for a 2-line to 4-line decoder

This VHDL program may now be applied to a synthesis tool to automatically generate a circuit for us! The actual circuit produced will be dependent on the type of device or technology (gate array, PLD, FPGA, etc.) but one implementation using simple gates is shown in Figure 1-11. You may want to verify for yourself that it implements the decoder function correctly.

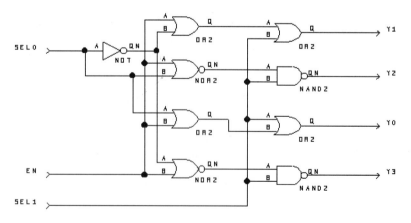

Figure 1-11: Circuit implementation of decoder using VHDL synthesis

This chapter has provided an introduction to the use of computer-aided engineering tools to develop and analyze digital logic circuits. Once you have completed the design exercises and tutorials in this book, you will be able to apply the skills you develop in a variety of different ways, whether you design using gates at the low level or design complete systems using one of the newer languages or tools available.

Good luck and have fun designing!

TUTORIAL INTRODUCTION TO *WORKVIEW OFFICE* TOOLS—*VIEWDRAW*, *VIEWSIM*, AND *VIEWTRACE*

CHAPTER
2

Introduction to
Workview Office
Student Edition

What Is Workview Office?

The *Workview Office* electronic design automation (EDA) software for Windows™ NT, Windows 95, and Windows 3.1x offers designers a complete suite of tools with the capability and high performance previously reserved for the UNIX platform. In addition to a full collection of traditional EDA tools, such as schematic entry, simulation, and waveform analysis, the *Workview Office* suite includes the industry's premier design tools — Vantage's SpeedWave™ VHDL simulator, Chronologic Simulation's VCS™ Verilog simulator, along with MOTIVE™ static timing analysis and XTK™ signal integrity tools from Quad Design.

Workview Office Student Edition

The *Workview Office Student Edition* release contains a subset of the *Workview Office* tools:

- **Project Manager** — adds, selects, and modifies design projects.

- **Library List Editor** — modifies the project search path of libraries and directories.

- **ViewDraw for Windows** — schematic editor, draws schematics and symbols.

- **ViewSim for Windows** —gate level simulator, simulates digital designs and back-annotates simulation values in *ViewDraw* designs.

21

- **Navigator for Windows** — browses and locates objects (nets, components, pins, etc.) in *ViewDraw* designs.

- **ViewTrace** — waveform analyzer, displays simulation results in waveforms for analysis and back-annotates values in *ViewDraw* designs.

- **Utilities** — prepares, accesses, and generates schematic netlist information.

The tools are arranged for easy access in the *Workview Office* toolbar (Windows 95) or a program group (Windows 3.1). The toolbar can be placed in a movable window as shown in Figure 2-1.

Figure 2-1: Workview Office toolbar in movable window

Restrictions and Limitations

The *Workview Office Student Edition* release tools have certain restrictions compared to the normal release of the product. There are two major differences. First, the maximum number of modules that can be used in a complete design is 300. A module is a primitive component such as an AND gate or a NOR gate. Notice that a module is not necessarily the same as a device. For example, a '161 four-bit binary counter is actually represented internally by approximately 30 modules. You will not be allowed to save a design that has more than 300 modules.

The second limitation is a restriction on the design names that can be used. Twenty design names are provided. Ten of these are to support the laboratory exercises in this book:

FULLADD.1	FOURBIT.1
ALU.1	LOOKAHD.1
RANDOM.1	TRAFFIC.1
TRAFFIC2.1	LOCK.1
VENDING.1	ARITH.1

The other 10 are provided for any other designs you want to implement:

DESIGN0.1	DESIGN1.1
DESIGN2.1	DESIGN3.1
DESIGN4.1	DESIGN5.1
DESIGN6.1	DESIGN7.1
DESIGN8.1	DESIGN9.1

Notice that all the design names must end with the extension .1.

3 Installation and Quickstart

System Requirements

Supported Operating Systems
- Windows NT V3.51
- Windows 95
- Windows 3.1 or later

80486/66MHz
16 MB RAM
40 MB of available disk space
800×600 video resolution
CD-ROM

Installation

The following procedure provides a few simple steps for installing the *Workview Office Student Edition* product. Select the installation procedure for your particular operating system.

To Install *Workview Office* Under Windows 95

1. Place the *Workview Office Student Version* CD into the CD-ROM drive of your PC.

2. Follow the Installation Wizard.

> **NOTE:** If the Installation Wizard does not start up within a minute or so, please follow the Windows NT 3.51 installation instructions shown below, but select the **Run** function from the Windows 95 Start menu.

3. Turn to the *Installation Process* section on page 25.

To Install *Workview Office* Under Windows NT 3.51

1. Place the *Workview Office Student Version* CD into the CD-ROM drive of your PC.

2. Select **File** ⇒ **Run** from the Windows Program Manager.

3. Enter `<cd_drive>:\WVOINST.EXE` where `<cd_drive>` is the letter of your CD drive.

4. Click **OK.**

5. Follow the Installation Wizard.

6. Turn to the *Installation Process* section on the next page.

To Install *Workview Office* under Windows 3.1x

To install *Workview Office Student Edition* from the CD-ROM for **WINDOWS 3.1** you must locate the correct file on the CD-ROM. This is contained in the WVO712 directory.

> **IMPORTANT:** In your CONFIG.SYS file ensure that the FILES = value is set to at least 50. For example, FILES = 50.

1. Place the *Workview Office Student Version* CD into the CD-ROM drive of your PC.

2. Select **File** ⇒ **Run** from the Windows Program Manager.

3. Enter `<cd_drive>:\WVO712\WVOINST.EXE` where `<cd_drive>` is the letter of your CD drive as shown in the following figure (E was the letter of the CD drive in this case):

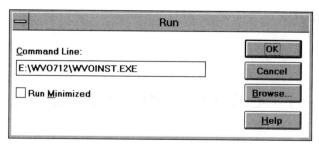

Figure 3-1: Selecting the correct install file

4.   Click **OK.**

5.   Follow the Installation Wizard.

> **IMPORTANT:** During installation *Workview Office* will check the WIN32S files on your computer and report if they are missing or out of date. If necessary follow the instructions to install WIN32S. Reboot after installing WIN32S and start the installation of *Workview Office* again.

6.   Turn to the *Installation Process* section.

After installation is complete the *Workview Office Program Group* is created:

Figure 3-2: Workview Office program group

The Installation Process

During the installation process you will be asked to provide some information. In many cases you can simply accept the default configuration.

When the installation process asks you for your name and company name enter your name and the name of your educational institution.

The *Workview Office* files and executables will be installed in the destination directory (assuming you accepted the default installation directory):

```
C:\WVOFFICE
```

At the end of the installation you can have the *autoexec.bat* file automatically updated to include the following required lines:

```
PATH=C:\WVOFFICE;%PATH%

SET WDIR=C:\WVOFFICE\STANDARD
```

Alternatively, you can manually modify the *autoexec.bat* file yourself.

When the installation is finished remember to reboot your system to set the environment variables before using *Workview Office.*

To Install the Libraries

During the installation process two libraries of components (*Builtin* and *74LS*) are copied from the CD-ROM to the following directory:

```
C:\WVOFFICE\LIBS
```

The resulting directory is shown in Figure 3-3.

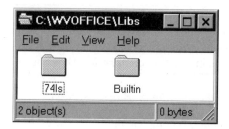

Figure 3-3: 74LS and Builtin libraries added

NOTE: Additional 74 series simulation libraries are contained on the CD-ROM. They are located in directory WVLIBS on the CD-ROM. These libraries are not required for the tutorials or lab exercises contained in this book. However, they may be copied (using File Manager or Windows Explorer) to a directory for your use in other designs.

Quickstart—If You Really Can't Wait!

After the *Workview Office Student Edition* has been installed (see previous sections for installation instructions) it is recommended that you start the Schematic Entry tutorial in the next chapter to learn how to create a new project and start a design using *ViewDraw*.

If you are familiar with using *Workview Office* and want to start using the tools immediately, make sure that you setup the *Project* appropriately before you start your design (see the first part of the tutorial if you are unfamiliar with these steps).

Also remember that the *Student Edition* release has name restrictions on the design names (see page 22 for the restrictions and limitations).

Schematic Entry Tutorial Using *ViewDraw*: Design of a Four-Bit Adder

This chapter introduces you to the process of creating a schematic or circuit diagram of a digital logic system. We use the four-bit adder design from Laboratory Exercise 2 (see Chapter 9) for an example. The Viewlogic schematic entry tool is called *ViewDraw* and through a sequence of simple exercises arranged in the form of a tutorial we show how to design the complete four-bit adder digital circuit.

> **NOTE:** If you are designing a different logic circuit than the four-bit adder, you can still use this tutorial to introduce you to designing with *Workview Office*. The only differences will be in the specific logic gates and circuits you will be working with.

In the following chapter we show how to verify that our four-bit adder works correctly through simulation using *ViewSim* and *ViewTrace*. The tutorials in this chapter and the next are to teach you the minimum steps necessary to complete the design-simulate-analyze process of creating a digital design in *Workview Office*. You will learn to use the following tools:

- **Workview Office Project Manager**—creates, selects, and removes project directories and libraries for a particular project.

- **ViewDraw**—schematic and symbol designer.

- **ViewSim**—gate level digital simulator.

- **ViewTrace**—waveform analyzer.

Before we start with the actual tutorial we provide an overview of the actual design methodology that we will be using.

Hierarchical and Flat Design Representations

There are a variety of ways to design a four-bit binary adder. Different devices can be used to implement the required functionality, and there are different ways to actually draw and represent the design. We could assume we had a very large piece of drawing paper and place on it all the logic gates and wires required to implement the design. This type of design is called *flat*, in that the whole circuit can be viewed at one level. Or, we could split up the design and place it on multiple smaller sheets and show the connectivity between sheets by labeling wires going from one sheet to another—but we would still have a flat design, although the smaller sheets would be easier to carry around and file.

An alternative, and usually better, methodology is to create a hierarchical design. In this approach the design is decomposed into a number of smaller functional blocks. At the top level of the design only the overall structure of these blocks would be shown, while the detailed design information would be available at lower levels in the hierarchy. Systems represented in a hierarchical manner are usually easier to create, modify, and understand.

We will draw our four-bit adder as a hierarchical design.

At the lowest level we will create a design of a single full adder as shown in Figure 4-1.

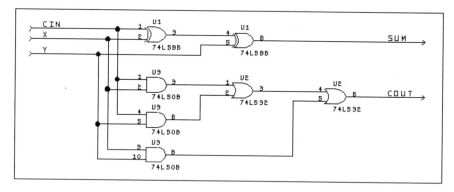

Figure 4-1: Schematic of a full adder

Because our four-bit adder requires four copies of this full adder, we can create a symbol representation of this diagram which we can then replicate four times. Figure 4-2 shows the symbol created to represent the full adder circuit.

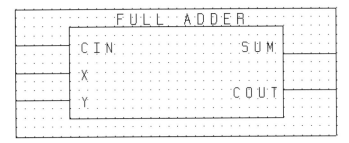

Figure 4-2: Symbol of the full adder

Figure 4-3 shows the completed design of the four-bit adder using four of the full-adder symbols. This sheet would be considered the top level in our design hierarchy. It allows the functionality of the system to be readily determined. If more detail on a particular aspect of the system design is required then the lower level, more detailed, schematics can be reviewed.

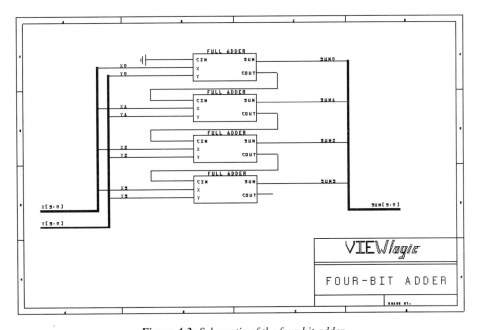

Figure 4-3: Schematic of the four-bit adder

Assumptions

Before you start the tutorial, complete the following tasks:

- Install *Workview Office Student Edition* on your system–for example, in C:\WVOFFICE as described in Chapter 3, *Installation and Quickstart*.

- Copy the *Builtin* and *74LS* library directories to C:\WVOFFICE\LIBS. These two libraries are copied automatically during the installation process. (To use an alternative 74 series library you will need to copy this from the CD-ROM.)

- Edit your AUTOEXEC.BAT file to reflect the proper search path.

> **NOTE:** During *Workview Office* installation, you can choose either to edit the AUTOEXEC.BAT file manually or to let the *Workview Office* installation do it for you.

After you install *Workview Office Student Edition*, and each time you boot your system, the AUTOEXEC.BAT file automatically sets up your *Workview Office* environment. The following list shows the variables that must be set (the *Workview Office* installation can do this for you automatically):

- PATH - sets your overall search path to the location of the *Workview Office* executable programs and batch files:

  ```
  SET PATH=C:\WVOFFICE;%PATH%
  ```

- SET WDIR - sets the *Workview Office* search path:

  ```
  SET WDIR=C:\WVOFFICE\STANDARD
  ```

The WDIR search path points to directories that contain configuration (.CFG) files, menu (.MN) files, macro (.MAC) files, initialization (.INI) files, and other associated files. The installed files reside in the \WVOFFICE\STANDARD directory, although you can copy them to another directory and modify them. If you do copy them to another directory, include that directory first in the WDIR path; for example,

```
SET WDIR=C:\NEW_STD;C:\WVOFFICE\STANDARD
```

The Workview Office Taskbar

The *Workview Office* installation creates a taskbar (or program group in Windows 3.1) containing all the installed products, as shown in Figure 4-4.

Figure 4-4: The Workview Office taskbar

You can run the *Project Manager*, *ViewDraw*, and other tools, directly from this taskbar in Windows 95, or from the *Workview Office* program group in Windows 3.1.

In Windows 95, by clicking on the left icon (the *Workview Office* icon) a menu of commands is available, as shown in Figure 4-5.

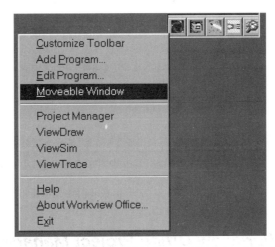

Figure 4-5: *Commands available from the Workview Office icon*

It is recommended that you change the fixed taskbar to a movable window as shown in Figure 4-6. This provides more flexibility in placing the toolbar.

Figure 4-6: *Movable window showing the Workview Office toolbar*

If you place the mouse pointer on one of the toolbar icons, the name of the tool will be displayed, as shown in Figure 4-7 (the mouse pointer was placed on the second left icon):

Figure 4-7: *Displaying the name of the toolbar icons*

Starting the Tutorial

About This Tutorial

This section contains a number of exercises to start you on the process of designing a four-bit binary adder (Laboratory Exercise 2 from Chapter 9) with the *Workview Office* schematic entry drawing package called *ViewDraw*. In this part of the tutorial you will learn about:

- Using the *Workview Office Project Manager* to add a project and add libraries to a project's search path

- Opening a new schematic sheet

- Adding a sheet border component

- Adding a component symbol to start a four-bit adder schematic design

Later sections of the tutorial show how to complete the design.

Using the Workview Office Project Manager

Opening the Workview Office Project Manager

In the following procedure, you will create a project using the *Workview Office Project Manager*.

Workview Office uses the concept of a *project* to contain and manage information regarding your designs. It automatically creates directories into which your schematics, symbols, and design details will be saved, and it contains files that you can use to customize the *Workview Office* environment.

The *Project Manager* is used to create and edit *Workview Office* project files. The project file *<file>.vpj* contains the project specific information, including the primary directory and the library search order.

In the following procedure we create a new project. We have called the project LABS and created a primary directory called C:\LABS to correspond to the laboratory exercises that we will be working on in this book. You can call your project something else and set up your primary directory in a different location. For example, you could use something like C:\EE3801\HOMEWORK, where EE3801 represents the course number of your digital logic class.

1. Invoke the *Project Manager* from the *Workview Office* taskbar or program group by clicking on the *Project Manager* icon ▣.

2. The *Workview Office Project Manager* dialog box opens:

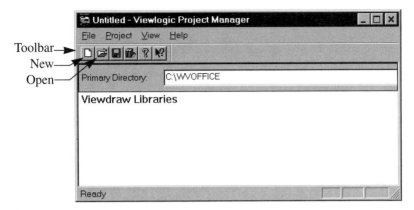

Figure 4-8: Workview Office Project Manager

NOTE: In the following descriptions and figures it is assumed that *Workview Office* was installed on drive **C:** in location **C:\WVOFFICE**. If you installed *Workview Office* on another drive or a different location then just substitute your new path in the examples.

HINT: Point the mouse at one of the toolbar commands and leave it there. After a short delay a hint box will appear showing the name of the command and a brief description will appear at the bottom of the window.

3.   Change the name of the *Primary Directory* field to define a new directory called LABS by typing in C:\LABS, as shown in Figure 4-9.

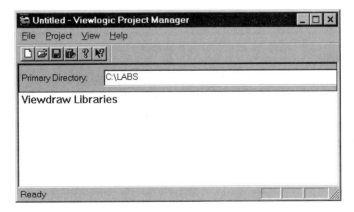

Figure 4-9: Changing the Primary Directory to C:\LABS

We will use this directory as the main directory for all the laboratory exercises.

When you create a new project the project manager automatically does a number of things. First, a project directory is created containing **\PKT**, **\SCH**, **\SYM**, and **\WIR** subdirectories (if they do not already exist):

- **SCH**—holds files that contain graphical descriptions of your schematics
- **SYM**—does the same for symbols
- **WIR**—contains logical descriptions of schematics

The *Workview Office Project Manager* also copies the **VIEWDRAW.INI** file, which contains startup information for *ViewDraw* from **C:\WVOFFICE\STANDARD** to the new project directory.

A project directory can contain a number of different designs consisting of schematics and symbols related in some way. For example, we have created a project called **LABS** to contain all the designs related to the lab exercises in this book. If you were going to work on another set of designs for another course or for project work then you would probably want to create a new project to contain those designs.

Although you can create many project directories, you can use only one project at a time.

Library Search Order

In your design you will be placing various types of components, such as logic gates, graphic modules, and symbols that you have created, onto your schematic. These components reside in a number of libraries and before you can access them you have to provide *Workview Office* with a list of the libraries you want to use and their directory location.

The following procedure adds libraries and directories to the current project's search path.

With *Workview Office Student Edition* two libraries are automatically copied from the CD-ROM. One is called *Builtin* and contains some general symbols for use in your schematics. The *Builtin* library also contains information that is required for simulating your designs. The second library supplied with the Student Edition is *74LS* and contains an extensive library of 74LS parts. We use both of these libraries for the tutorials and laboratory exercises in this book.

> **NOTE:** You should add only libraries and directories that you are going to use. *ViewDraw* searches through the *entire* list of *Current Libraries* in the *Library Search Order* each time you open a design (symbol or schematic) or add a component.

1. In the *Project Manager* window select the **Project ⇒ Libraries** menu command.

   The *Library Search Order* dialog box opens:

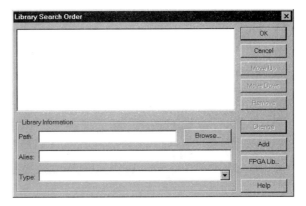

Figure 4-10: Library Search Order dialog box

We need to add the *Builtin* and the *74LS* libraries to the library search order.

The following procedure adds the *Builtin* library.

Adding the Builtin Library

This procedure adds the Workview Office *Builtin* library, which is required for digital simulation. The *Builtin* library includes *ViewSim* primitives from which you can create customized simulation models. The *Builtin* library also contains a number of graphical items (like a sheet border) and other components such as ground and power symbols.

1. In the Library Information *Path* field, you can *either* enter the full path of the *Builtin* library, or click on **Browse** to find the required library directory path.

    If you click on **Browse**, the *Select Directory* dialog box is displayed:

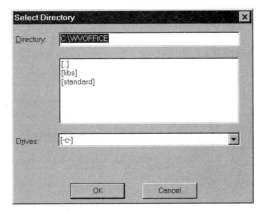

Figure 4-11: Library List Editor Select Directory dialog box

2. Either type C:\WVOFFICE\LIBS\BUILTIN in the Directory field or double-click on *libs* and then double-click on the *builtin* directory to enter it in the *Directory* field.

    Figure 4-12 shows the *Select Directory* dialog box after you select the *builtin* directory:

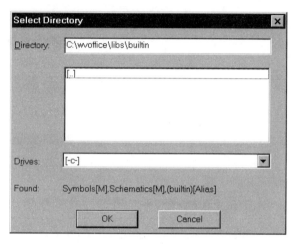

Figure 4-12: Selecting the Builtin library

3. Click on **OK** to return to the *Library Search Order.*

4. Click on **Add**. The *Library Search Order* now looks like this:

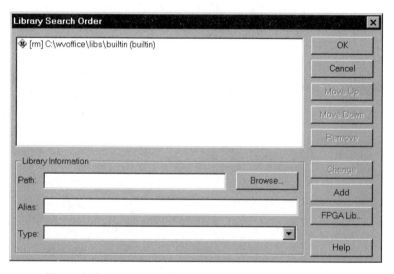

Figure 4-13: Library List Editor after adding the Builtin directory

Adding the 74LS Library

In the same way you added the *Builtin* library, add the *74LS* library (**C:\WVOFFICE \LIBS\74LS**) to the library list. The sequence is

1. Select **Browse.**

2. In the *Select Directory* dialog box enter the C:\WVOFFICE\LIBS\74LS direc-
    tory, as shown in Figure 4-14.

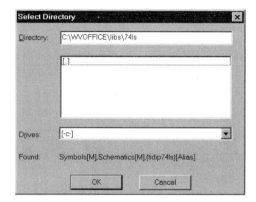

Figure 4-14: Selecting the C:\WVOFFICE\LIBS\74LS directory

3. Click on **OK**.

    You return to the *Library Search Order* dialog box.

4. In the *Library Search Order* window click on **Add** to add the *74LS* library, as in
    Figure 4-15.

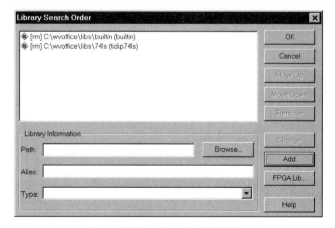

Figure 4-15: Adding the 74LS library

5. Click on **OK.**

You return to the *Project Manager* dialog box as shown in Figure 4-16.

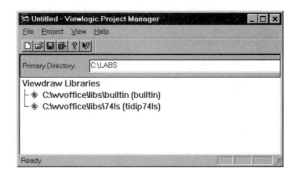

Figure 4-16: *Project Manager showing the two libraries*

6. Select the **File** ⇒ **Save As** menu command.

A dialog box will open confirming that you want to create the new primary directory C:\LABS (Figure 4-17).

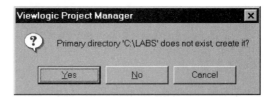

Figure 4-17: *Creating the new LABS directory structure*

Click on **Yes.**

7. Type *labs* for the file name in the *Save As* dialog box:

Figure 4-18: *Entering the LABS project file name*

8. Click on **Save.**

This save procedure will create the *labs.vpj* file used to define the project.

9. Select the **File** ⇒ **Exit** menu command.

NOTE: By default, the active project will be the last .vpj file that was loaded by the *Workview Office Project Manager*. We will use the same project for all the designs in this book.

Some Additional Notes on Libraries and Aliases

The Alias Field—In the *Library Search Order*, you may have noticed a name in parentheses after the *Builtin* and *74LS* names. This name is an alias for the complete directory path. Aliases for Viewlogic-shipped libraries are specified in the LIB.CFG file in each library directory; these aliases are automatically included in the project's search path.

NOTE: Aliases uniquely identify symbols of the same name from different libraries. For example, (mylib)74LS500 and (TI) 74LS500 use the same symbol name but come from different libraries.

IMPORTANT! Aliases must include parentheses and begin with an alphabetical character.

There are four library types:

Primary—You must specify a primary directory where *ViewDraw* stores all user-created symbols and schematics.

Read Only—Specifies a library to which you have read-only privileges.

Read/Write—Specifies a directory to which you have read and write privileges.

Megafile—Specifies directories in which symbol, schematic, and wirelist files are stored together, rather than in separate files, to save disk space.

The *Library Search Order* automatically sets the default library type to Read Only; it assigns a Megafile library type to Viewlogic-shipped libraries.

Starting the Design

Starting *ViewDraw*

The first step in starting a design is to create a new schematic in *ViewDraw* and add a sheet border component—This provides a frame or boundary for your design.

1. Click on the *ViewDraw* icon on the *Workview Office* customized toolbar (or select *ViewDraw* from the *Workview Office* menu command).

    The following dialog box opens:

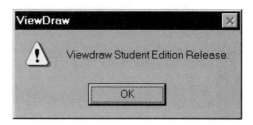

Figure 4-19: ViewDraw Student Edition dialog box

2. Click on **OK**.

3. The *ViewDraw* window opens:

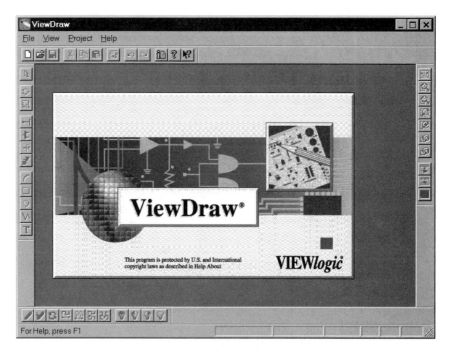

Figure 4-20: ViewDraw

The *ViewDraw* logo will disappear after a few seconds or you can click on it to close it.

    ViewDraw uses a number of toolbars for frequently used commands and operations. The *standard toolbar* is located at the top of the *ViewDraw* window just under the menu bar and includes quick mouse access to commands such as save, cut, copy, paste, undo, redo, and help. The *object toolbar* is located on the left hand side of the design window and provides access to frequently used commands in *ViewDraw*. Down the right hand side of the design window is the *view toolbar*, and along the bottom is the *transform toolbar*. Finally the *status* bar is located at the bottom of the *ViewDraw* window and includes information such as sheet number and current mode. We will use these toolbars as we create our design.

ViewDraw can be used to draw both schematics and symbols. We will first create a schematic of our full adder and then create a symbol for the full adder that we can use later in the design of the complete four-bit adder.

When a project is created, a primary directory is created along with a set of subdirectories. The subdirectories and other files for our LABS project are shown in Figure 4-21.

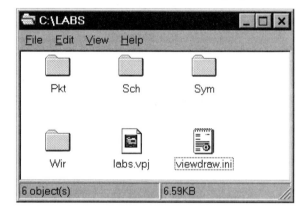

Figure 4-21: Directory structure of the LABS project

The *labs.vpj* file contains the project information and the *viewdraw.ini* file contains configuration information for *ViewDraw* and also lists the project libraries.

The Sch subdirectory will be used to store the schematics for our design while the Sym subdirectory will store the symbols.

We can now start the design of our full adder schematic.

Creating a New Schematic Sheet

1. Select the **File** ⇒ **New** menu or toolbar command.

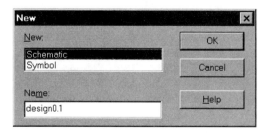

Figure 4-22: Creating a new schematic design

2.   Make sure Schematic is selected.

3.   In the *Design Name* field, enter the name **fulladd.1** (make sure you add the .1 extension) and press **Return** (or click on **OK**):

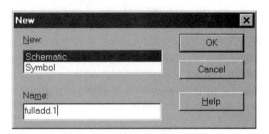

Figure 4-23: Starting a new schematic called fulladd

IMPORTANT

In the *Workview Office Student Edition* the names of the designs are limited. There are 10 names to support the laboratory exercises in this book:

```
fulladd, fourbit, alu, lookahd, random,
traffic, traffic2, lock, vending, arith.
```

There are also 10 names to support your other designs:

```
design0, design1, design2, design3, design4,
design5, design6, design7, design8, design9.
```

NOTE: All the design names must have the extension .1, which indicates that this is sheet number one of a design.

If you enter a different name, for example, test1, you will get an error message:

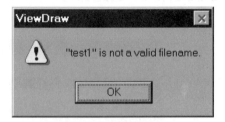

Figure 4-24: Invalid filename warning

If you enter a valid design name, a default B-size sheet displays in the window, as shown in Figure 4-25. The design name is in the header—\sch\fulladd (notice that this design is a schematic and will be stored in the SCH subdirectory).

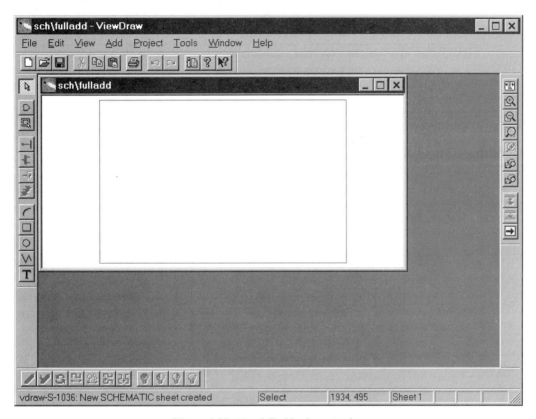

Figure 4-25: *New fulladd schematic sheet*

Also notice the *standard toolbar* that runs across the top of the design window just under the menu bar. This toolbar offers easy access to some general commands like opening and saving designs, cutting and pasting, printing, and accessing help.

Down the left hand side of the design window is the *object toolbar*. Like the horizontal toolbar, it offers easy mouse access to many commands such as adding components, nets, or text. On the right side is the *view toolbar* which provides zoom commands and commands for moving up and down sheets in a design hierarchy. Along the bottom is the *transform toolbar* which can be used for deleting, rotating, and stretching components.

The toolbars are explained in more detail later in the tutorial.

You can hide the toolbars using the toolbar commands in the **View** menu command.

> **HINT:** You can find a short description of each of these commands by resting the mouse pointer on the icon for a few seconds. Or you can click on the Help button 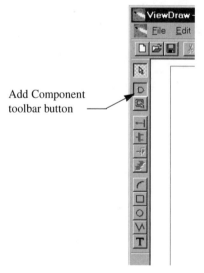 in the toolbar, position the help pointer on an item and click to obtain a detailed description.

> **HINT:** You may want to view your schematic at full size. First make your *ViewDraw* window full size by clicking the maximize button of the *ViewDraw* window. Then make your schematic window full size by clicking the maximize button of the schematic window. Finally press **F4** to show a full view of your schematic.

Adding a Sheet Border

We now need to start adding components and other items to our schematic by selecting the *Add Component* command.

1. There are a number of different ways to select the *Add Component* command.

    *   Choose **Add** ⇒ **Component** from the pull down menu or click the *Component* button on the object toolbar as shown in Figure 4-26. (Alternatively you can type the keyboard shortcut key 'c').

Add Component
toolbar button ——

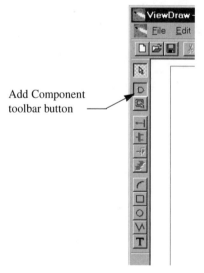

Figure 4-26: *Selecting the component command from the object toolbar*

After the *Add Component* command has been selected by one of the above methods, the *Add Component* dialog box is displayed:

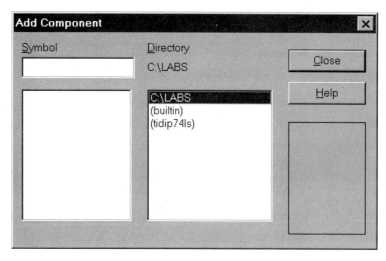

Figure 4-27: Add Component dialog box

2. In the Libraries list, click on the **(Builtin)** directory.

3. Scroll down the symbol list in the left column to find BSHEET.

4. Click on **BSHEET.1** in the Components list. The BSHEET symbol appears in the *symbol preview* window in the lower right corner of the dialog box.

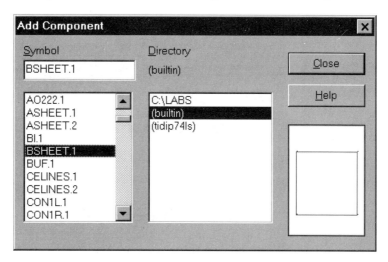

Figure 4-28: Selecting the BSHEET component in the Builtin library

5. Drag the component from either the component list or the symbol preview window onto the schematic sheet. The sheet border component appears as shown in Figure 4-29.

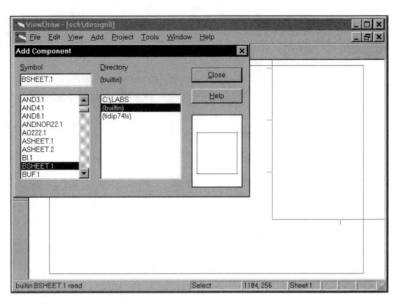

Figure 4-29: Dragging the BSHEET component onto the schematic

6. Move the sheet border component with the mouse to position it properly against the actual sheet boundary.

7. Release the left mouse button to place the component on the schematic.

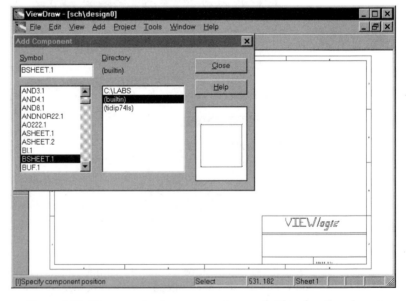

Figure 4-30: The screen border component correctly placed on the schematic

> **HINT:** If you misplace the border component, you can drag it to its correct place with the mouse pointer.
>
> Note that to move a component it first has to be selected by clicking on it or by pressing the left mouse button and dragging a selection box around the component.
>
> The screen border component is unusual in that the selection point is in the bottom left corner. To select it, hold down the left mouse button while you drag a box around the bottom left corner of the sheet. Once the component is selected you will see a small box displayed. A $number also appears which is the system identifier for this particular component.
>
> Once the small box is displayed you can point at this with the mouse cursor and by holding down the left mouse button drag the border to the correct place.
>
> You can also use the **Edit** $\Rightarrow$ **Undo** command (or the undo button on the standard toolbar), then re-execute the add component command to re-add the component.

8. Click the **Close** button to dismiss the *Add Component* dialog box.

Adding Text to the Schematic

This section shows how to add text to your schematic. It is important to add descriptions to your designs to explain to other engineers what your circuit is and how it works. At the very least you should provide a title for the design with the name of the engineer responsible for the design (you!) and the date the design was drawn or last modified.

We can now add the title "Full Adder" to the schematic by selecting the *Add Text* command.

1. First, deselect the BSHEET component by clicking anywhere in the middle of the sheet.

2. Select the *Add Text* command.

    There are a number of ways to do this.

    The quickest way is to click on the *Add Text* button in the vertical object toolbar, as shown in Figure 4-31.

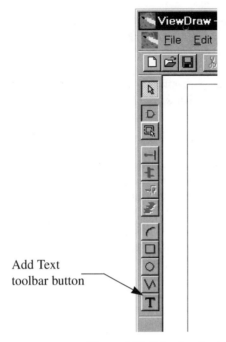

Add Text
toolbar button

Figure 4-31: Selecting the Add Text icon

Alternatively, the *Add Text* command may be selected by the **Add** ⇒ **Text** menu command or the **Shift + T** keyboard shortcut can be used.

3.   Click and hold the left mouse button in the middle of the schematic.

     A small text string appears.

4.   Drag the mouse to move the text string to the box underneath the word "View-logic" in the lower right corner of the schematic.

     Release the left mouse button.

     The *Text Properties* dialog box displays:

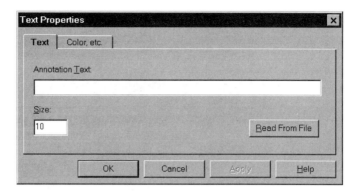

Figure 4-32: Add Text dialog box

5. Enter **Full-Adder.**

6. Click on the *Size* field and enter **34** to enlarge the text; click on **OK**.

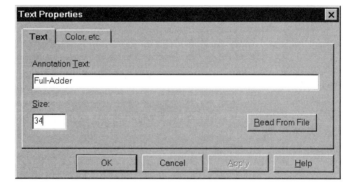

Figure 4-33: Adding the Full-Adder text and changing the size

When you click on **OK**, you will see the text appear on the screen.

7. Clear the *Add Text* mode by clicking on the *Select Mode* button at the top of the vertical object toolbar.

8. By clicking and holding down the left mouse button on the "Full Adder" text,
    drag the text underneath the word "Viewlogic" in the lower right corner of the
    schematic, as shown in Figure 4-34.

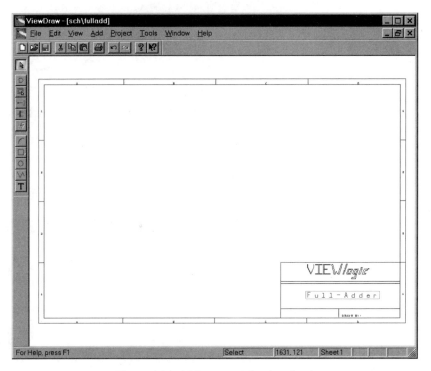

Figure 4-34: *Adding text to the sheet border*

9. Release the left mouse button to place the text.

    The red bounding box around the text indicates that this item is currently
    selected. You can deselect an item by clicking elsewhere in the schematic.

Adding a Component

You have just created and labeled a schematic sheet and added a sheet border. In this section, we start adding logic gates to the schematic.

Procedure

1. Choose the **Add** ⇒ **Component** menu command or press the *Add Component* toolbar button to display the *Add Component* dialog box.

2. Click on the **(tidip74ls)** library, then scroll through the components and click on **08.1** (the .1, .2, .3, etc., extensions all provide different representations of the same component).

    (Or, you can type **08** in the *Symbol* box.)

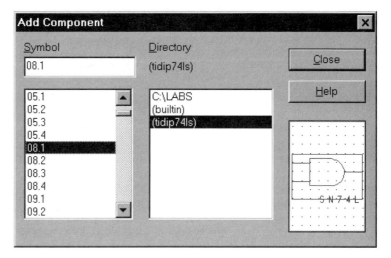

Figure 4-35: Selecting the 74LS08 component

3. Drag the AND gate from the list or the lower right viewer screen onto your schematic to the approximate location shown in Figure 4-36.

    > **HINT:** You may want to move the *Add Component* dialog box to another area of the screen before dragging the component.

    > **NOTE:** Use only integrated circuits from the *74LS* library (alias **tidip74ls**). Do not select gates from the *Builtin* library because these do not represent physical devices and so do not have the necessary attributes for timing analysis.

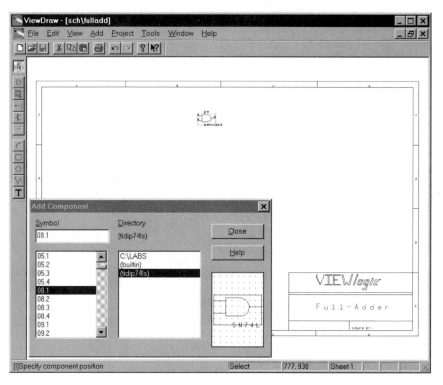

Figure 4-36: Placing the 74LS08 component

4.   Click anywhere in the window to deselect the object.

Saving Your Schematic

You can save your schematic at any time during the design process.

The horizontal toolbar provides a number of icons related to saving and working with your design files.

> **HINT:** If you place the cursor over one of the icons and leave it there, a message will be displayed in the message area at the bottom of the screen describing the action of that icon.

1.   To save your file, select the *file save* icon from the toolbar or the **File** ⇒ **Save** menu command.

2.   You can also check your schematic with the **File** ⇒ **Save + Check** menu command.

> **HINT:** After you select **File** ⇒ **Save + Check**, a dialog box appears to display any warnings about the schematic, such as an unconnected component. This is a typical error that *ViewDraw* looks for when it saves a schematic design. It is perfectly acceptable to save a schematic with errors, the file is still saved. However, these errors would have to be corrected before the design could be used for simulation.

3. Click on **CLOSE** to acknowledge that you verified this warning.

4. Select **File** ⇒ **Exit** to close the schematic and quit *ViewDraw*, or select **File** ⇒ **Close** to close the schematic but keep *ViewDraw* active.

Adding More Components

We now need to add some more components to our full-adder design.

> **NOTE:** If you quit *ViewDraw* in the previous section, invoke *ViewDraw* and select **File** ⇒ **Open** to open the dialog box and double-click on **FULLADD.1** to display it in the *ViewDraw* window.

From the **tidip74ls** library, add a 74LS86A component above the 74LS08 component you placed earlier and a 74LS32 component to the right of the 74LS08 component. Your schematic should now look similar to Figure 4-37.

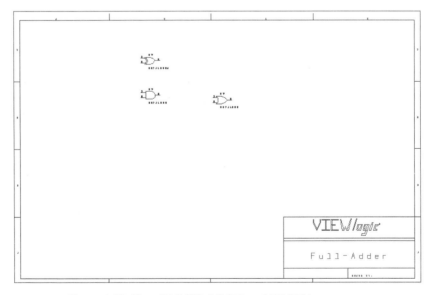

Figure 4-37: *Placed 74LS08, 74LS32, and 74LS86A components*

We now need to add the remaining components to complete our design. Before we do this you may find it easier if you zoom in on the area of the schematic that you want to work in. The following procedure shows you how to do this.

Zooming in on an Area of the Schematic

1. Move the cursor to the upper left corner of the schematic and press the **F9** key.

2. While holding down the left mouse button, drag the mouse diagonally from the upper left to the lower right (you can drag the select box diagonally up or down in any direction), as shown in Figure 4-38, to create a select box, then release to zoom in on the area inside the box.

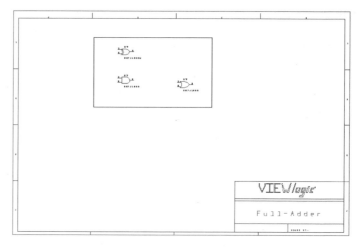

Figure 4-38: Defining a zoom area

Figure 4-39 shows the result of the zoom. (Depending on how far you zoom you may not see the grid dots.)

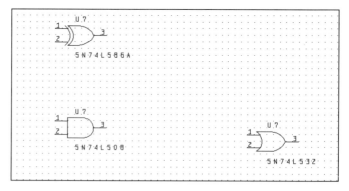

Figure 4-39: Zoom results

The above command was the zoom region command. There are also function keys and menu commands to zoom in and out and display full views of the design:

1. Select the *Zoom In* button ![icon] from the view toolbar. (Alternatively select the **View** ⇒ **In** menu command or press **F7**). You zoom in more. Repeat again.

2. Now select the *Zoom Out* button ![icon] or select the **View** ⇒ **Out** menu command or press **F8.** You zoomed back out to view the components.

3. The *Zoom Full* button ![icon] or the **View** ⇒ **Full** menu command or **F4** displays the whole schematic maximized to fit the design window.

Adding More Components

We can now add the remaining components by selecting **Add** ⇒ **Component** or pressing the *Add Component* toolbar icon in the same way we added the first three gates. This is probably the quickest way, but an alternative method is to use the **Copy** and **Paste** command. These commands are available from the **Edit** menu or from the toolbar, as shown in Figure 4-40. We will use the copy and paste commands to copy the 74LS86A gate.

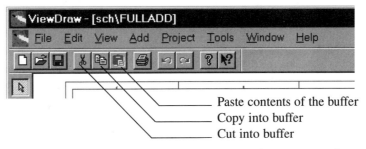

Figure 4-40: Toolbar icons for buffer cut, copy, and paste commands

1. First select the 74LS86A component. Move the cursor over the component and select it by clicking on it.

2. Select the *Copy into Buffer* icon or the **Edit** ⇒ **Copy** menu command.

    This copies the selected component into the buffer.

3. Select the *Buffer Paste* icon or the **Edit** ⇒ **Paste** menu command.

4. Click in an empty part of the schematic.

    A new copy of the component appears at the cursor.

5. Move the component by dragging it with the mouse (press and hold the left mouse button) to the required location for the second 74LS86A component. Release to place the component, as shown in Figure 4-41.

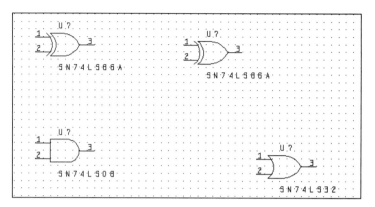

Figure 4-41: Placing a second 74LS86A gate by copying and pasting

Because we used the copy and paste command, we have obtained a copy of the first gate. Notice that the pin numbers for the second gate are the same as the first gate – 1, 2, and 3. However a 74LS86A integrated circuit has four EXCLUSIVE-OR gates, each using unique pin numbers. The first gate occupies slot number 1 and uses pins 1, 2, and 3. The second gate occupies slot number 2, and uses pins 4, 5, and 6.

We can change the slot position of the copied gate as follows:

6. Select the **Edit** ⇒ **Slot** menu command.

7. Change the default slot position from 1 to 2, as shown in Figure 4-42.

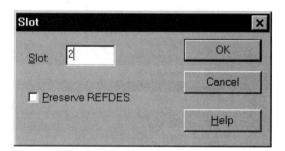

Figure 4-42: Changing the component slot position

8. Click on **OK**.

    Notice that the second gate from the 74LS86A integrated circuit is now shown—denoted by the pins numbered 4, 5, and 6.

9. In a similar manner, copy two more 74LS08 gates and one more 74LS32 gate and place them in the approximate positions shown in Figure 4-43. Use the **Edit** ⇒ **Slot** menu command to change the slot positions of the additional gates.

> **Remember:** To move a component, first select it by clicking on it. Then move it by dragging it with the mouse (press and hold down the left mouse button) to the required location.

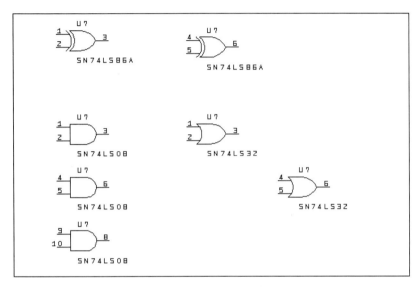

Figure 4-43: Adding the remaining components for the full adder

> **NOTE:** Instead of adding additional components by copying and pasting, you can use the *Add Component* command (selected from the toolbar on the left of the *ViewDraw* window or from the **Add ⇒ Component** pull-down menu command) as described earlier. In this case, the correct slots for the components are automatically assigned. Try it! Delete the third 74LS08 gate (by clicking on it and pressing the **DEL** key) and add a new 74LS08 gate using the add component command. The pins for the new gate will correspond to slot three—pins 8, 9, and 10.

In the next section, you learn how to add wire connections (nets) between components.

Adding Nets and Buses

Overview

After you place components on a schematic, you connect them with *nets* and *buses*.

A *net* represents an electrical connection with one or more segments.

A *bus* is a collection of nets that can operate as a group or as individual nets within buses. A bus groups related signals. For example, the bus that you create in this chapter represents several signals.

You specify net and bus names with labels, which you will create in the following section, *Labeling*.

This section describes the following topics:

- Adding a net between two pins

- Adding dangling nets that you will connect later

- Making electrical connections between component instances with nets (wires)

Adding a Net Between Two Pins

If you quit *ViewDraw* in the previous section, invoke *ViewDraw* and select **File** ⇒ **Open** to open your schematic and double-click on **FULLADD.1** to display it in the *ViewDraw* window.

1. Zoom in on the area you want to work in by moving the cursor to the upper corner of the schematic. Press the **F9** key, drag the mouse diagonally over the area you want to enlarge, and then release. Also remember the zoom in and zoom out commands (**F7** and **F8** or the view toolbar commands).

2. Select the *Add Net* mode by selecting the **Add** ⇒ **Net** menu command, or click on the *net* icon in the vertical toolbar as shown in Figure 4-44.

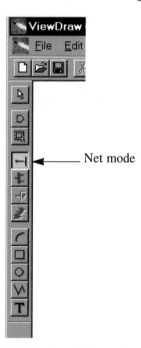

◄——— Net mode

Figure 4-44: Selecting the Net mode

This activates the *Add Net* mode in *ViewDraw*.

Notice that the current mode is displayed in a box at the bottom of the *View-Draw* window.

3. Move the mouse cursor to pin 3 of the 74LS86A, and click and hold the left mouse button.

4. Drag the net to pin 4 on the second 74LS86A component and release when you reach pin 4.

> **HINT:** If you accidentally connect the net to the wrong pin, select **Edit ⇒ Undo** to delete that net.

You should now see the net shown in Figure 4-45:

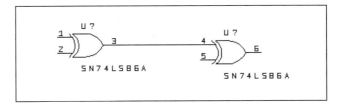

Figure 4-45: Adding a net

Adding Dangling Nets

A dangling net ends in a square box—it does not connect to anything (as yet). You will add nets and buses to these dangling nets later in the tutorial.

1. The *Add Net* mode should still be active from the last operation. If not, select it again.

2. Move the mouse cursor to pin 1 on the 74LS86A and drag the net left a short distance and release.

    A dangling net (shown as a net with a square on the end) is created.

    The only time you should see a dangling net is when just one end is connected to a component, bus, or net. If you have a net between two objects and still have a square (nonconnection), click on the box to continue wiring the net.

3. Do the same thing to pins 2, 5, and 6 on the two 74LS86A components as follows:

    Move the cursor to pin 2, click to start drawing the net. Drag the net out to the left and release to complete the net. Do the same thing for pin 5. For this net we need to route it around the first EXCLUSIVE-OR gate. Note that routing around another component is automatically carried out—but if you want to force a 90-degree turn at a certain place then just click at the required point. Finally, add the dangling net for pin 6. Your schematic should look similar to Figure 4-46.

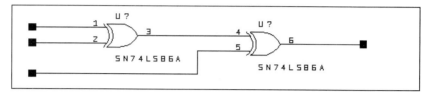

Figure 4-46: Adding dangling nets

> **HINT:** Use F6 to pan or move across a schematic. Wherever you place the cursor becomes the center point of your view. Point the cursor where you want the center of your view to be, and press **F6**.

We now need to add the remaining nets (wires) to complete the design of the full adder. Before we do this it might be useful to look at the **Move** command. You can move individual components, wires, or other items or move a group of objects. You can experiment by moving the gates (either in isolation or with nets attached) on your design.

Moving Components

To move a component it must first be selected.

1. Clear the *Add Net* mode by selecting the *Select Mode* icon ![icon] at the top left of the vertical toolbar.

2. Select the second EXCLUSIVE-OR gate by placing the cursor on it and click and hold the mouse button and move the cursor to move the gate to the required new position.

3. Release when you reach the new position.

4. If any of the nets need altering after the move you can delete them (**Edit ⇒ Cut** menu command or simply press the **DEL** key after selecting them) or move them in a similar way to moving a component.

    You may want to use this opportunity to move some of your other gates in preparation for adding the remaining nets.

Adding More Nets

We now connect one of the AND gates to the already placed nets.

1. Ensure that *add net* mode is activated (the current mode is displayed at the bottom of the *ViewDraw* window). Select the **Add ⇒ Net** menu command or select the *Add Net* vertical toolbar icon.

2. Move the cursor to pin 1 of the 74LS08 and click and drag the net a short distance to the left. Then drag the net upward to reach the top already placed net, and release.

3. Connect a net to pin 2 of the AND gate in a similar manner. Your schematic should look similar to Figure 4-47.

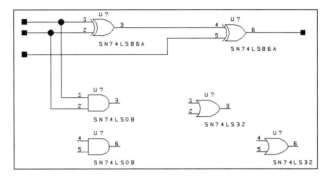

Figure 4-47: Adding more nets

4. Continue adding the remaining nets to complete your design. Your schematic should look similar to Figure 4-48.

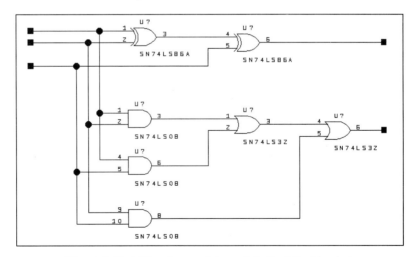

Figure 4-48: Adding the remaining nets to the full-adder design

NOTE: If you make a mistake, remember you can use the **Edit ⇒ Undo** menu command or press the *Undo* toolbar icon to cancel an action.

5.   Press the *File Save* icon or the **File ⇒ Save + Check** menu command to save and check your schematic.

Labeling

Labeling is the process of identifying a net or component by assigning it a text string. Labels are important. They are used to maintain connectivity from one part of the design to another and are also used during simulation to apply and monitor signal values. In this session, you learn how to label the nets in your full-adder design.

Labeling Multiple Nets

1.   Zoom in on the upper left corner so that you can clearly see the three nets.

2.   Double-click on the net connected to pin 1 of the EXCLUSIVE-OR gate.

    The *Net Properties* dialog box displays.

    We could label the nets one at a time. However, to save time we can provide one list of all the labels as described in the next step.

3.   In the Label entry field, enter the following (notice there are no spaces between the names):

      CIN,X,Y,SUM,COUT

    The *Net Properties* dialog box should look as follows:

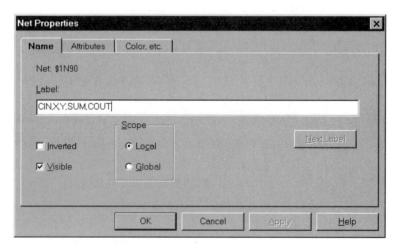

Figure 4-49: Entering the net labels in the Net Properties dialog box

4.   Press **Enter** or click on **OK**.

The CIN label appears at the end of the net (do not worry about the exact placement of the label for the moment)

5. Double-click on the next net (X) to select it.

   The *Net Properties* dialog box displays.

6. Click on the **Next Label.**

   The X label appears in the *Label* field.

7. Press **Enter** or click on **OK.**

   The X label appears at the end of the net.

8. Double-click on the third net (Y) to select it and display the *Net Properties* dialog box.

9. Click on the **Next Label.**

   The Y label appears in the *Label* field.

10. Press **Enter** or click on **OK.**

11. Do the same operations to label the two nets on the right of your diagram (SUM and COUT).

    We now need to move the labels to improve the appearance of the schematic.

12. First, clear the *Add Net* mode by selecting the *Select Mode* icon. To move a label just drag it with the mouse (be careful not to select the net as well or you will move both).

    The resulting schematic should look similar to Figure 4-50.

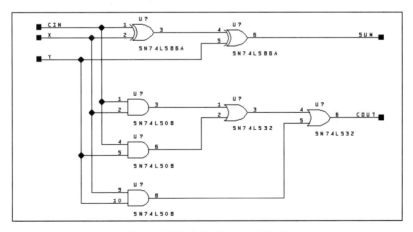

Figure 4-50: Labeling the full adder

> **HINT:** To change an existing label or to change the label size, double-click on the label. The *Label Properties* dialog box will display.

Finishing the Design

Overview

In this section, you learn how to do the following:

- Add inputs and outputs to the schematic

- Save and check the schematic for errors

Adding More Components

We need to add input and output ports to the dangling nets in our full-adder schematic.

1. Select the **Add** ⇒ **Component** menu command or the *Add Component* icon from the toolbar.

2. From the *Builtin* library select the IN.1 component.

3. Drag the component with the mouse to correctly place it just to the left of the CIN dangling net, then click.

4. The dangling net will disappear, as shown in Figure 4-51.

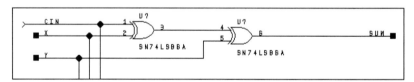

Figure 4-51: Adding an IN component

5. Either drag a new copy of the IN component or use the **Edit** ⇒ **Copy** menu or toolbar command and the paste command to display a copy of the IN component.

6. Drag the new copy of the IN component and place it on the X net.

7. Do the same for another copy of the IN component and place it on the Y net.

8. In a similar manner, add OUT.1 components (also from the *Builtin* library) to the SUM and COUT nets.

9. Your final schematic should look like Figure 4-52.

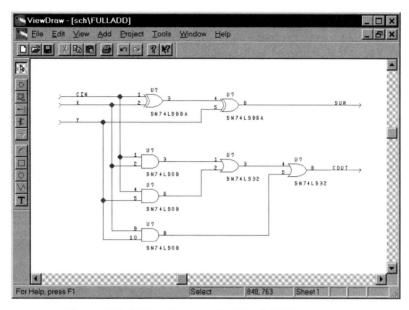

Figure 4-52: *Adding the remaining IN and OUT components*

Saving and Checking the Schematic

To save your schematic, you can use the **File** ⟹ **Save** menu command or the *save* toolbar button at any time. However, this **Save** command does not check the logic in your design. To both *save* and *check* your schematic for logical errors, use the **File** ⟹ **Save + Check** menu command.

NOTE: You must use the **File** ⟹ **Save + Check** command before simulating the design in *ViewSim*.

---

1. Select the **File** ⟹ **Save + Check** menu command.

    If your schematic is error-free, a status message at the bottom of the *ViewDraw* window confirms that there are no errors or warnings.

    HINT: If you mislabeled nets or buses, left any nets out, or did not properly connect nets, the *ViewDraw Message Tracker* dialog box displays errors and warnings.

    If you know what the error is, click on **Close**, correct any errors, and save and check the design before proceeding with the tutorial.

    You can also use the **Visit** button to highlight the area where any problem exists.

    For example, if you had a component that was not connected to the rest of your schematic you would get a message similar to that shown in Figure 4-53.

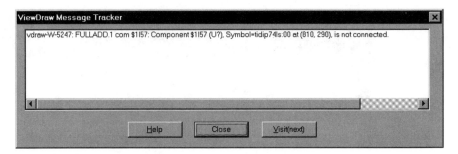

Figure 4-53: Displaying ViewDraw errors and warnings

2.   Select the **File** ⇒ **Close** command.

When you save a schematic, *ViewDraw* also creates a wirelist file (an ASCII file that describes the logical connectivity of your design) from the schematic and places it in the current project's **\WIR** subdirectory.

The **vsm** utility, described in a later section, uses the wirelist file to create a **.VSM** network file for input to *ViewSim*.

We have now completed the design of our full adder schematic. To make the complete four-bit adder we need four of these full adders. We will create a symbol representing the full adder that we can then replicate four times for the final four-bit design.

Creating Symbols

In this section, we create a *composite symbol*, which includes an underlying schematic, for the full adder. We then use this as a component to create the complete four-bit binary adder.

Create a User-Defined Border

1.   In *ViewDraw*, select the **File** ⇒ **New** command.

The *New* dialog box appears.

2.   Select **Symbol** (not schematic) and then enter **FULLADD.1** in the *Name* field.

3.   Click on **OK.**

A large square appears. This is the outer boundary of the symbol.

4.   Double-click inside the square.

The *Symbol Properties* dialog box appears:

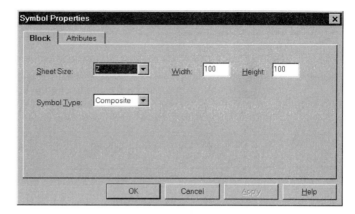

Figure 4-54: Default symbol properties

5. Change the width size from 100 to **360**. (The units are in 1/100ths inch and so entering 360 produces a block with a width of 3.6 inches.)

6. Change the height from 100 to **140** (for a block height of 1.4 inches) and press **Return**. (Press **F4** to display a full view of the symbol)

    The border changes (notice that the header indicates the name of the design):

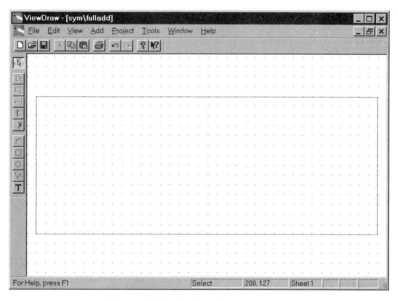

Figure 4-55: The boundary for the symbol

Adding a Symbol Body

When you create the body of any type of symbol, you graphically construct it using commands in the vertical toolbar menu.

1. Select the **Add** ⇒ **Box** menu command or the *Box* command from the vertical toolbar on the left side of the *ViewDraw* screen.

2. Create a body inside the symbol boundary as shown in Figure 4-56 by clicking six grid squares right and two down from the top left corner, dragging the box down to the bottom right and releasing to end the box.

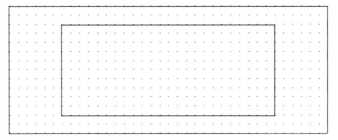

Figure 4-56: Drawing a body

In the next procedure, you add a pin to the symbol. A pin represents a port or interface on the symbol. The pins directly relate to the *in* and *out* nets of the full adder schematic we completed earlier.

Adding a Pin

NOTE: Do not use lines to create pins. Use the **Add** ⇒ **Pin** menu command or *pin* toolbar command.

1. Select the **Add** ⇒ **Pin** menu command or the *Pin* command from the vertical toolbar.

2. Move the cursor near the top left of the body.

3. Click and drag the cursor left to the boundary box and release.

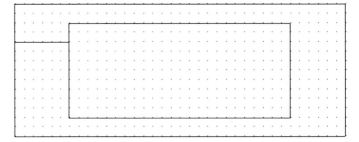

Figure 4-57: Adding a pin

4. Adding the remaining pins:

Repeat the above process to add the remaining pins as shown in Figure 4-58.

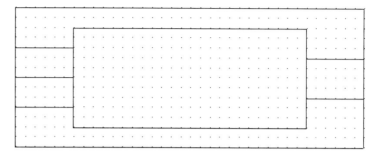

Figure 4-58: Adding the remaining pins

Adding a Pin Label

1. Clear the *Add a Pin* mode by clicking on the *Select Mode* icon (the arrow) at the top of the vertical toolbar.

2. Double-click on the top left pin.

   The *Pin Properties* dialog box displays.

3. In the label field, enter **CIN** as shown in Figure 4-59.

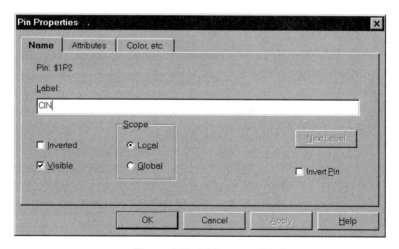

Figure 4-59: Adding a pin label

4. Click on **OK** or press **Return**.

5. Making sure that only the label is selected (click once on the CIN label then click and hold the mouse pointer on the CIN label to move it), move the label to the approximate position shown in Figure 4-60.

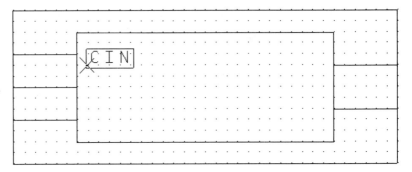

Figure 4-60: Labeling the CIN pin

> **NOTE:** To allow the labels to be aligned better with the pins it is possible to change the grid size. Use the **Project ⇒ Settings** command and change the Grid Spacing from 10 to 5.
>
>   You may have to zoom in to see the grid dots.
>
>   You can also change the label origin point (default is bottom left corner) with the **Project ⇒ Settings** command and change the Text Origin (try middle center).

Adding More Labels

1. We can now add the remaining labels to the pins. We label the remaining pins using the *Expand Label* option.

2. Double-click on the left middle pin (X pin).

   The *Pin Properties* dialog box displays.

3. In the label field, enter **X,Y,SUM,COUT** (notice there are no spaces) and press **Return** or click on **OK**.

   An *Expand Label* query box displays.

4. Click on **YES**.

   The label is placed on the pin (you can move it later).

5. Repeat for the remaining pins, as follows:

   Double-click to select a pin, press the **Next Label** button, press **OK**.

   After placing and moving the remaining labels, your symbol should look similar to Figure 4-61.

NOTE: If you ever see remaining bits of graphics after a move or other operation pressing **F5** will refresh the screen.

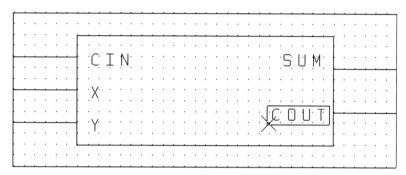

Figure 4-61: Adding the remaining labels

Adding Text to the Symbol

You can add text to give the symbol a name that displays when you place an instance of it on a schematic. Text is not attached to an element like a label is—text only provides information about the symbol on the schematic.

1. Select the **Add** ⇒ **Text** menu command or select the *Text Mode* toolbar icon.

   Click and hold the left mouse button near the top of the symbol. A small text string appears.

2. Release the left mouse button.

   The *Text Properties* dialog box displays.

3. In the annotation text field, enter **FULL ADDER** (text is case-sensitive; labels always appear in uppercase).

4. Change the size to 15 and change the color to RED.

5. Click on **OK**.

6. Deselect the text mode by clicking on the *Select Mode* icon ⬜ from the object toolbar.

7. By clicking and holding down the left mouse button, drag the "FULL ADDER" text to the position shown in Figure 4-62.

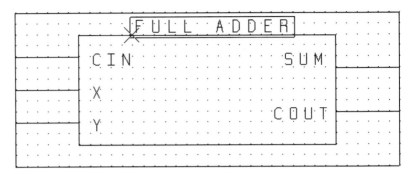

Figure 4-62: Placing text on a symbol

HINT: You can double-click on the text item to display the *Text Properties* dialog box to change the size, color, etc.

8. Click in an empty area of the window to deselect the text.

Adding Symbol Attributes

Attached and Unattached Attributes—In symbol drawings, you can have attached and unattached attributes.

Attached attributes apply only to a particular pin—for example, the pin number (#) or PINTYPE attribute. You add attached attributes to a symbol pin by selecting the pin first and then adding attributes.

Unattached attributes apply to the whole symbol; you do not select anything before adding them to the symbol.

Attribute Visibility—You can make all attributes invisible, except the pin number (#) and REFDES attributes. The values of these are always displayed, regardless of the visibility setting. The reason is that these attribute values change for each instance of the component on your schematic; the other attributes remain the same.

HINT: Making the attributes invisible does not delete them; it just helps eliminate clutter from the schematic display.

Attaching Pin Attributes

1. Double-click to select the CIN pin.

2. The *Pin Properties* dialog box opens.

3. Select the *Attributes* page.

4. In the *Name* field, enter PINTYPE.

5. Tab to the *Value* field and enter IN as shown in Figure 4-63.

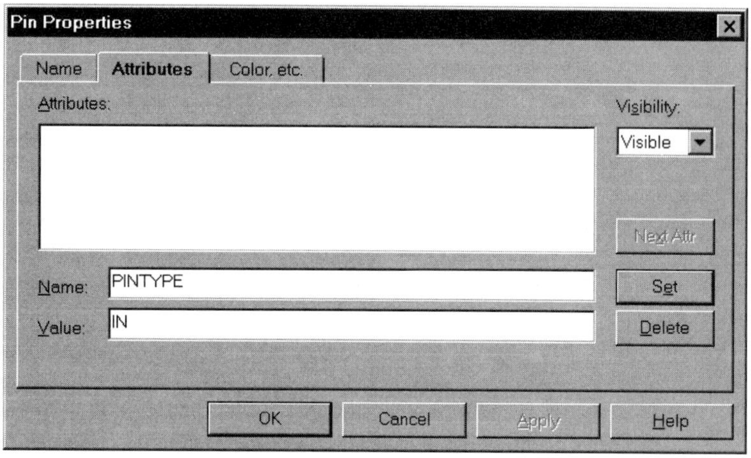

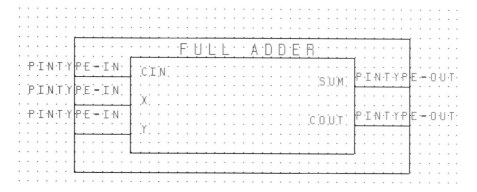

Figure 4-63: Adding pin attributes

6. Click on **OK** to return to the symbol. It doesn't matter where the attribute PINTYPE = IN appears because we will be making it invisible later on.

   Repeat the above process to add PINTYPE = IN attribute to the X and Y pins and to add PINTYPE = OUT attribute to the SUM and COUT pins.

7. Your symbol should look similar to Figure 4-64.

Figure 4-64: Adding the remaining PINTYPE attributes

8. We now need to make the PINTYPE attributes invisible to reduce the visual clutter on the schematic. We can do this in one of two ways:

   Select the PINTYPE attributes by clicking once on them and then select the *Make Invisible* command from the transform toolbar at the bottom of the *ViewDraw* window, or

Double-click on each of the PINTYPE attributes and change the visibility from visible to invisible.

9. Press **F5** to refresh the screen if necessary.

> **HINT:** You can deselect an object by moving the cursor to an open area of the *ViewDraw* window and clicking.

Saving the Symbol

1. Double-click outside the symbol.

   The *Symbol Properties* dialog box opens.

   Notice that the Symbol Type is *Composite*.

   This indicates that the symbol will represent a base function and will include an underlying schematic (the default symbol type is composite).

2. Select **OK** or **CANCEL**.

3. Select the **File ⇒ Save** command.

   Notice the window displaying information about your symbol—scroll through to see the symbol attributes. Close the window (**File ⇒ Close)**.

   > **NOTE:** You have already created the underlying schematic (FULLADD) with the same name as your symbol so *ViewDraw* knows that it is the underlying schematic for that symbol.

4. Select the **File ⇒ Close** to close the symbol.

Designing a Four-Bit Adder

So far you have created a one-bit full adder schematic and a corresponding symbol. We now want to use this symbol to create a four-bit binary adder circuit. Because we have created a symbol of the full adder we can now add copies (called instantiations) of it to new schematics just like adding any other component.

We open a new schematic called **FOURBIT**, add four copies of the full-adder component, interconnect them, and finally add input and output buses to complete our design.

Creating a Four-Bit Adder Schematic

1. Select **File ⇒ New** command.

   Enter **FOURBIT.1** in the *Design Name* field (make sure **Schematic** type is selected).

A new schematic sheet displays in the window.

We now need to add a sheet border.

2. Select **Add** ⇒ **Component** menu command or *Component* mode from the vertical toolbar, select the *Builtin* directory, select the BSHEET.1 in the symbol list and drag it to place the sheet border (review *Adding a Sheet Border* on page 44 at the start of this tutorial if you need help).

3. Using the **Add** ⇒ **Text** menu command or *Text* mode, add the name **Four-Bit Adder** to the schematic (review *Adding Text to the Schematic* on page 47)

4. Select the *Component* mode again.

    Ensure that the LABS library is selected (you should see a FULLADD.1 component in the symbol list).

    We now add one instance of the previously created FULLADD component.

5. Click on the FULLADD.1 name—Notice that the symbol appears in the preview window.

6. Drag it to the upper part of the schematic as shown in Figure 4-65.

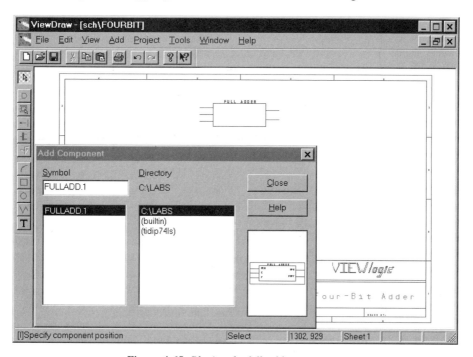

Figure 4-65: *Placing the full-adder component*

We need to place four copies of our previously designed full adder symbol to create a four-bit adder. We could place each of the four full adder symbols one by one using the **Add** $\Rightarrow$ **Component** command, but we introduce a new command called **Add** $\Rightarrow$ **Array** to show how to place multiple copies of components on a schematic.

First let us add some nets to the full adder.

Select the *Net* mode toolbar command, and attach dangling nets to the full-adder component as shown in Figure 4-66.

Figure 4-66: Adding nets to the full-adder component

Creating an Array of Components

You can create multiple instances of objects using a component array instead of adding the same object over and over again.

1. Make sure the *Add Net* mode is off (press the *Select Mode* toolbar icon ![icon] if necessary)

2. By starting in the upper left, hold down and drag the box that appears so that you select the entire full adder component and its attached nets.

3. Release to select the group.

4. Select the *Add Array* command ![icon] from the object toolbar or select the **Add** $\Rightarrow$ **Array** menu command.

    The *Array* dialog box appears.

5. Change *Rows* from 1 to 4.

    Positive numbers create the array to the right and/or upward in the window.

6. Change *Row Spacing* from 10 to -1.

    The negative number creates an array downward in the window.

7. Change the *Spacing* from Absolute to Relative.

    Relative Spacing indicates that the column and row spacing will be measured from the border of the minimum bounding box that surrounds the selected object(s).

Absolute Spacing indicates that the column and row spacing will be measured from the lower left corner of the minimum bounding box that surrounds the selected object(s).

8. Your dialog box should look as follows:

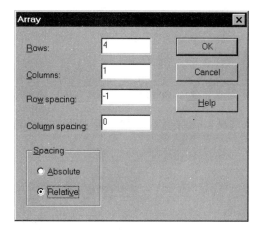

Figure 4-67: Creating an array of components

9. Click on **OK**.

The objects that you selected are now repeated three more times down the schematic, as shown in Figure 4-68.

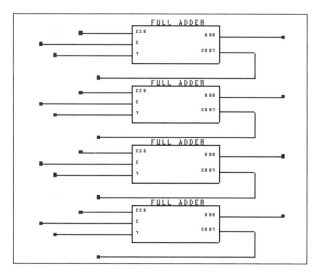

Figure 4-68: Adding an array of components

Connecting the Full-Adder Components

We now need to connect the COUT pin of one full adder to the CIN pin of the next full adder.

1.   Select the *Net* mode toolbar command.

2.   Point the cursor at the COUT dangling net on the top full adder component and click.

3.   Draw a net segment by dragging the mouse to the next full adder's CIN dangling net and release to complete the net segment.

     Repeat for the other two full adder components. Your schematic should look like Figure 4-69.

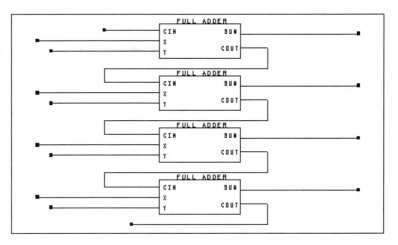

Figure 4-69: Connecting the COUT to CIN pins on the full-adder components

Adding Buses

A bus is a collection of nets that can operate as a group or as individual nets within buses. A bus groups related signals. We are going to create three buses; X, Y, and SUM, with each bus containing a group of four nets. These buses will help us later on when we want to apply binary numbers to the four-bit adder to confirm its correct operation through simulation.

1.   Select the **Add** ⇒ **Bus** menu command or press the *bus* mode toolbar icon.

2.   Click on the top right SUM dangling net's ending square.

3.   Drag the cursor down to about the position shown in Figure 4-70, connecting the other SUM dangling nets, release and click to add a bend, and continue to the right.

    4.   Release to end the bus.

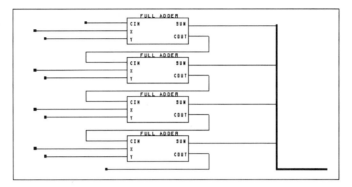

Figure 4-70: Adding the SUM bus

    5.   Repeat the same process to add the X and Y buses as shown in Figure 4-71.

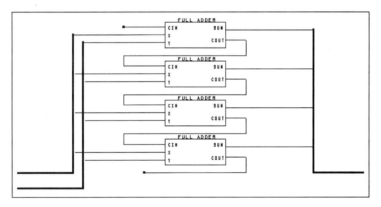

Figure 4-71: Adding the X and Y buses

    6.   Enter **File ⇒ Save + Check** to save and check your schematic.

> **NOTE:** You will receive bus label errors. Ignore them for now; you will label the bus and the nets in the next section.

Autoincrementing Labels

We could label the X, Y, and SUM buses and nets individually. But it is usually quicker to autoincrement the labels.

    1.   Clear the previous *Bus* mode.

2. Select the net connecting the bottom right SUM pin that goes from the bottom full adder component to the bus by double-clicking on it.

> **NOTE:** In this procedure where we are automatically incrementing the nets, the order of selection of the SUM nets is important. Make sure you select the SUM nets in the sequence SUM3, SUM2, SUM1 and SUM0, as shown in Figure 4-72.

The *Net Properties* dialog box displays.

3. Enter **SUM[3:0]** and press **Return**. A box appears on the net with SUM3 in it (don't worry where it is placed at the moment; we will move it later).

4. Move the cursor to the next SUM net above and double-click on it.

The *Net Properties* dialog box displays.

Select **Next Label**.

SUM2 appears in the *Label* field.

5. Click on **OK**.

6. Continue until the SUM1 and SUM0 nets are labeled as shown in Figure 4-72 below (you may want to move them from their default location to near the bus):

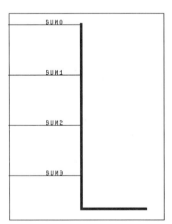

Figure 4-72: Placing the SUM[3:0] labels

Repeat the above process to label the X and Y nets, as shown in Figure 4-73.

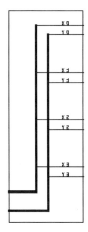

Figure 4-73: *Labeling the X and Y nets*

Labeling the Buses

1. Label the SUM bus segment by double-clicking on the far right bus segment to display the *Net Properties* dialog box.

2. Enter **SUM[3:0]** and press **Return** or click on **OK**.

3. Repeat the same process for the X and Y buses.

    Your schematic should now look like Figure 4-74.

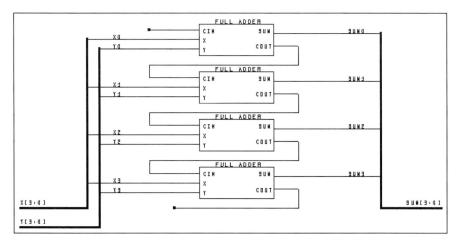

Figure 4-74: *X, Y, and SUM buses labeled*

4. Select **File** ⇒ **Save + Check** to save and check the **FOURBIT** schematic.

You should not have any schematic errors or warnings.

Finishing the Design

To complete the four-bit adder schematic we need to connect a ground to the CIN net of the first full adder (to apply a logic '0') and we also need to remove the dangling net from the COUT pin of the last full adder.

The dangling net connected to the COUT pin of the last full adder can be deleted by simply selecting it and then selecting the **Edit** ⇒ **Cut** menu command or by pressing the **DEL** keyboard key.

Adding a GND Component

1. Select the **Add** ⇒ **Component** menu command or *component* mode.

    Ensure that the *Builtin* library is selected.

2. Select the GND.1 symbol in the symbol list.

    Place it near the CIN net of the first full adder as shown in Figure 4-75.

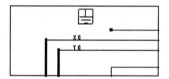

Figure 4-75: Adding the GND component

We now need to rotate the GND component.

3. Select the **Edit** ⇒ **Rotate** menu command or the rotate icon from the bottom toolbar.

4. Click on the GND component to turn it so it is pointing left (repeat steps 3 and 4).

5. Once the GND component is in the correct orientation, move the GND component so it connects to the dangling net on the CIN pin of the full adder, as shown in Figure 4-76.

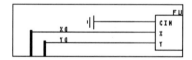

Figure 4-76: Final placement of the GND component

6. Select **File** ⇒ **Save + Check** to save your design and verify that there are no errors or warnings.

Viewing the Design

The following procedure shows you how to view a design in *ViewDraw* using the **View** menu, keyboard commands, and function keys to view areas of a design.

> **HINT:** Pressing **F5** refreshes the *ViewDraw* window.

1. Press **F6** to pan across the design. Click in the design to make that the center of the next view, then press **F6** again.

2. Press the *Zoom In* command from the view toolbar or press **F7** or select the **View ⇒ In** menu command to zoom into the design.

3. Press the *Zoom Out* command from the view toolbar or press **F8** or select the **View ⇒ Out** menu command to zoom out.

4. Press the *Zoom* command from the view toolbar or press **F9** or select the **View ⇒ Zoom** menu command to zoom a selected region.

   Drag the mouse to display a bounding box. Use this box to select the region of the design that you want to zoom in on.

   Release to zoom into the selected region.

5. Press the *Zoom Full* command or press **F4** or select the **View ⇒ Full** menu command to fit the design to the *ViewDraw* window.

6. Select a FULL ADDER symbol in the FOURBIT design.

7. Select the *Zoom Select* command or the **View ⇒ Selected** command to zoom in on the selected symbol as shown in Figure 4-77.

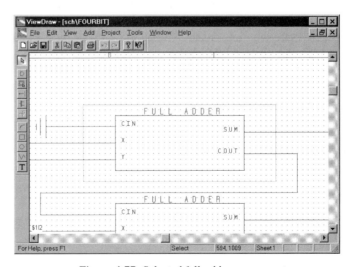

Figure 4-77: Selected full-adder component

8. Press the right mouse button.

    A pop-up menu appears.

9. If you select the **Schematic** menu option, the underlying fulladd schematic is displayed.

10. If you select the **Symbol** menu option, the underlying fulladd symbol is displayed.

11. Close down the new windows (**File** ⇒ **Close**) to return to the FOURBIT schematic design.

Creating a Network File

After you create and check a schematic in *ViewDraw*, the next step in the digital design process is to create a network file for input to *ViewSim*, the digital simulator for *Workview Office*.
You accomplish this step by running the design through the **vsm** utility.
Remember that *Workview Office* creates three directories in your project directory:

SCH—holds files that contain graphical descriptions of your schematics.

SYM—does the same for symbols.

WIR—contains logical descriptions of schematics.

The *ViewDraw* **File** ⇒ **Save + Check** command creates a file of your schematic in the \SCH area and runs the **check** utility. The check utility performs a connectivity check of your schematic and creates a WIR file. The WIR file contains complete connectivity and component attribute information about your design.
The *ViewSim* wirelister (**VSM**) utility uses the WIR file to extract the necessary information to support simulation. **VSM** creates a network file (with the extension .**VSM**) in the project directory.
The following procedure shows how to create this network file in *ViewDraw*.

Creating a Network File From *ViewDraw*

1. In *ViewDraw*, with your FOURBIT design open, select the **Tools** ⇒ **Create Digital Netlist** command, as shown in Figure 4-78.

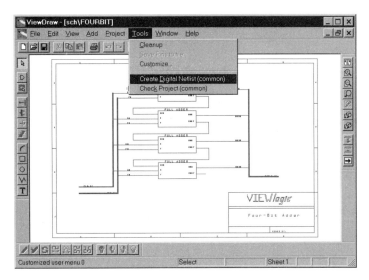

Figure 4-78: *Selecting the Create Digital Netlist command*

2. A DOS window will open and the *ViewSim Wirelister* will execute:

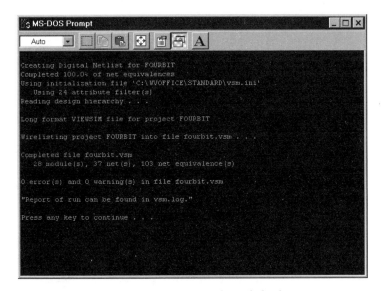

Figure 4-79: *ViewSim Wirelister dialog box*

Notice the information displayed. A file *fourbit.vsm* is created and there should be no errors or warnings.

Correct any error or warning messages before continuing. Once the design has no errors, **vsm** creates a **.VSM** file that is used by *ViewSim*.

NOTE: If you encounter warnings such as a schematic is newer than the wir
file open the *fulladd* schematic and *Save + Check*, then open the *fulladd* symbol and save it and finally *Save + Check* the *fourbit* schematic. Then run the
Tools ⇒ **Create Digital Netlist** command again.

3.   Close the DOS window by pressing any key.

5 Digital Analysis Tutorial Using *ViewSim* and *ViewTrace*

Using ViewSim and ViewTrace

This chapter describes how to simulate the FOURBIT binary adder design using *ViewSim*, review back-annotated values in *ViewDraw*, and analyze waveforms (simulation output) in *ViewTrace* by guiding you through the following steps:

- Opening the network file for the *ViewDraw* design (FOURBIT.VSM) that you created in *ViewDraw*

- Setting up a waveform stream between *ViewSim* and *ViewTrace*, based on selected nodes from *ViewDraw*

- Interactively executing *ViewSim* commands

- Simulating for a specified duration

- Viewing back-annotated simulation values on the *ViewDraw* hierarchical schematic

- Viewing simulation output as waveforms in *ViewTrace*

Starting *ViewSim*

1. Either from the *Workview Office* toolbar, or the *Workview Office* program group, start the *ViewSim* digital simulator.

A dialog box will open:

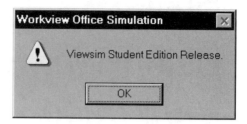

Figure 5-1: ViewSim tool starting

2. Click on **OK**.

ViewSim starts:

Notice the command line box—keyboard simulation commands are entered here.

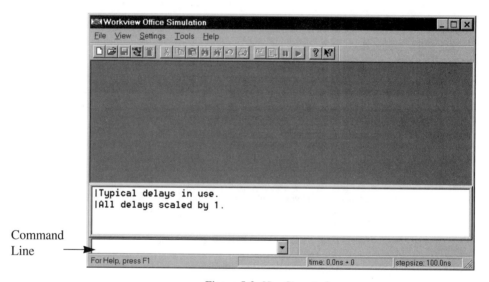

Figure 5-2: ViewSim window

Opening the Network (.VSM) File

After you create a network .VSM file from your *ViewDraw* schematic using the **Tools** ⇒ **Create Digital Netlist** command (which runs the **vsm** utility), you can open that network file for simulation in *ViewSim*.

During the simulation session keep the *ViewDraw* window open on your fourbit schematic design—we use this to show the values on the nets as the simulation progresses.

(At this point you should have the *ViewDraw* and *ViewSim* windows open.)

1. Click on the *ViewSim* window (to make it the active window).

2. Select the **File** ⇒ **Load ViewSim Netlist** command and select the FOURBIT.VSM network file that you created from your schematic, **fourbit.vsm,** as shown in Figure 5-3.

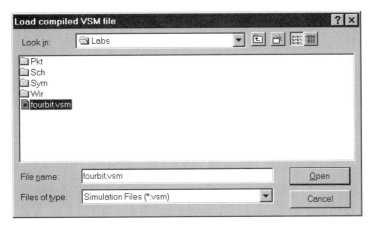

Figure 5-3: Selecting the fourbit.vsm netlist

3. Click on **Open**.

The fourbit.vsm netlist file will be loaded into *ViewSim*. Notice the message showing that your design had 28 modules.

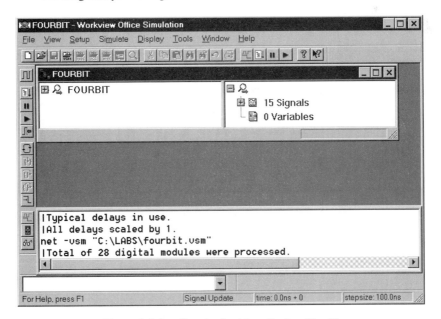

Figure 5-4: Loading the fourbit netlist into ViewSim

> **NOTE:** You can open only one network file (.VSM) at a time in *ViewSim*.
>
> **NOTE:** *ViewSim* reads the network information about the design and processes it for simulation. Several informational messages appear (see Figure 5-4).
>
> The phrase "digital modules" refers to parts in the Viewlogic *Builtin* library. Most standard parts and libraries from Viewlogic (and parts that you build) have *Builtin* primitives at the lowest level of hierarchy. (The 28 digital modules correspond to the four full adder components which each contain seven simple logic gates.)

Examining Simulation Values

After you invoke *ViewSim* and open a network file, back-annotated simulation values are displayed on the open schematic in the *ViewDraw* window. At this time, you have not issued any simulation commands. Therefore, these values default to the unknown state— X. Figure 5-5 shows the **FOURBIT** schematic; the elapsed Simulation Time displays in the lower right corner of the window.

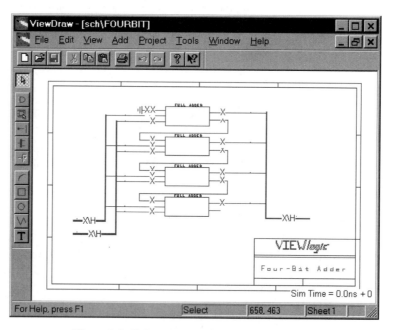

Figure 5-5: *Unknown states displayed in ViewDraw*

Starting the Simulation

Displaying Simulation Results in *ViewTrace*

We want to graphically display the results of our simulation using *ViewTrace*, the wave-form analyzer. To do this we need to establish communication between the simulator and *ViewTrace* by creating a waveform stream. The waveform stream is a temporary file created by the simulator and read by *ViewTrace*.

To create a waveform stream:

1. Select the **Setup** ⇒ **Waveform Streams** menu command. This opens the *Setup Waveforms Streams* dialog box:

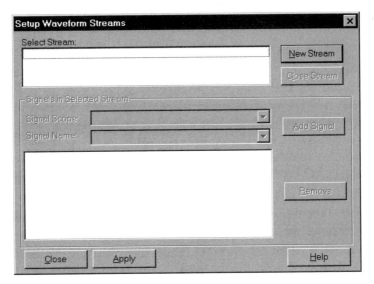

Figure 5-6: Setup Waveforms Streams dialog box

2. Click on the **New Stream** button.

   This opens the *Create New Waveform Stream* dialog box.

3. Ensure that the LABS directory is selected (navigate to this directory if necessary)—This directory contains the files we are simulating.

4. Enter **fourbit.wfm** in the *File Name* area and select Waveform Files in the *Save as type* area as shown in Figure 5-7.

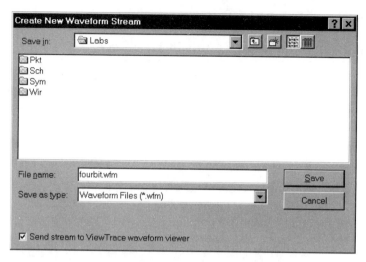

Figure 5-7: Creating the fourbit.wfm waveform stream

5. Click on **Save**.

    You return to the *Setup Waveform Streams* dialog box.

    We now need to identify the signals we are interested in graphically displaying.

    We display the X and Y inputs and the SUM outputs to show that our circuit correctly adds four-bit numbers.

    We can do this by entering the signal names directly in the *Signal Name* area of the dialog box. A quicker method is to use the Navigator to select the signals, as described in the following section.

The Navigator

The Navigator graphically displays the structure of any *ViewDraw* symbol or standalone VHDL model (we are not concerned with VHDL models). The Navigator presents the structure in a collapsible tree view with all the nodes in the hierarchy represented by an icon.

You can use the Navigator to traverse into any level of a design hierarchy and display the names and current values for signals at each level of the design.

1. Click on the *ViewSim* window to activate it.

2. Click on the **+** symbol near **FOURBIT** in the *Navigator* window in the *Simulation* window.

The Navigator displays the nodes of the fourbit design on the left—The four $numbers represent the four full-adder components in our design. The right pane lists the signals on the fourbit design. (Since this is not a VHDL model there will be no variables.)

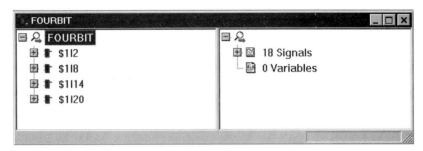

Figure 5-8: The Navigator window

1. Click on the **+** symbol near the **18 Signals** to list all 18 signals.

2. Scroll through the list. Notice the 15 signals, which we named (X0, X1, X2, X3, X[3:0], Y0, Y1, etc.), and the three other signals with the $number, which are signals on the schematic we did not name and which have the default $ label identifier.

3. Select all the X, Y, and SUM signals by dragging the mouse over these signals to select them. (You can press the **CTRL** key and the mouse to enable you to select multiple groups of signals at one time.)

4. Go back to the *Setup Waveform Streams* window.

5. Click in the *Signal Name* area.

6. Click **Add Signal** to add the signals.

7. Click on **Apply**.

ViewTrace starts and displays the following:

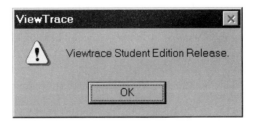

Figure 5-9: ViewTrace tool starting

The *ViewTrace* window opens, displaying the waveforms of the selected signals from your fourbit design as shown in Figure 5-10 (the window has been maximized to display all the signals).

Your *ViewTrace* window may have the signals in a different order.

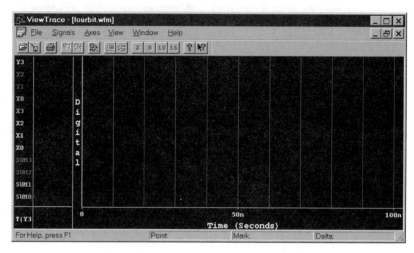

Figure 5-10: ViewTrace window showing waveforms of selected signals from fourbit

8.   Close the *Setup Waveform Streams* dialog box.

Setting Values and Starting Simulating

1.   To make it easier to assign values and manipulate our four-bit X, Y, and SUM buses, we will use the vector command.

Enter the following keyboard commands at the *ViewSim* command line (refer back to Figure 5-2 for the location of the command line on the *ViewSim* window) — Click in the command line box to activate it before typing commands.

> **NOTE:** Like most of the *ViewSim* (and *ViewDraw* and *ViewTrace*) commands, there are menu commands that are equivalent to keyboard commands. When simulating it is recommended that you use keyboard commands because it is usually quicker to type these in than using the pull-down menus. However, in the following examples the **Setup** ⇒ **Define Vector** or the **Setup** ⇒ **Stimulus** menu commands could also be used.

vector x x[3:0]

After entering the command the *ViewSim* window will look similar to Figure 5-11.

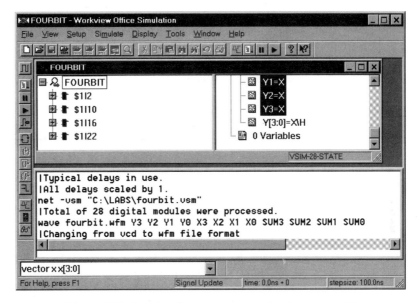

Figure 5-11: *Entering the vector command at the command line*

2. Press **Enter** on the keyboard to enter the command.

    This command assigns a vector called *x* to represent your *x* input bus. In a similar manner enter the following commands to assign vectors *y* and *sum* to represent your *y* input and *sum* output buses.

    vector y y[3:0]
    vector sum sum[3:0]

3. We can now initialize the *x* and *y* buses to zero (0) using the **assign** command (abbreviated to *a* in the following examples):

    a x 0\h

    This command sets or assigns the *x* vector (representing the *x* bus) to a value of 0 (zero). (The hexadecimal notation (\h) is strictly speaking unnecessary in this example.)

    Enter a similar command to set the *y* vector to 0:

    a y 0\h

4. These two commands will force the four bits of the *x* and *y* buses to be equal to 0. In the above two assignments the radix of the number was set to hexadecimal

by the use of the \h notation after the number). Values may be entered in hexa-decimal, decimal, octal, and binary (the default radix for a vector is binary).

> **NOTE:** Instead of setting all the fourbits in one command as shown above we could set individual bits using the force high or low commands. For example, the following two commands would set bit 2 of bus *x* to logic 0 and bit 3 of bus *y* to logic 1 (notice that the **low** command is abbreviated to *l*, and the **high** command is abbreviated to *h*):
>
>
>
> l x2
>
> h y3
>
>

5. Now that we have set up our initial input conditions to the four-bit adder, we can start the simulation by issuing the simulation (abbreviated to **sim**) command. With no time units or period specified to the **sim** command the default simulation is 100 ns (equal to the current *stepsize,* which by default is 100 ns).

    sim

    You also use the **Simulate** ⇒ **Run** or **Simulate** ⇒ **Run Stepsize** menu command or the *run* command from the vertical toolbar.

> **NOTE:** If after running a simulation for a certain period of time the circuit is not stable (indicated by a warning message), issue a further **sim** command. If the circuit is not stable it means the signals are still propagating through the gates.

Figure 5-12 shows the *ViewSim* window after entering the above commands.

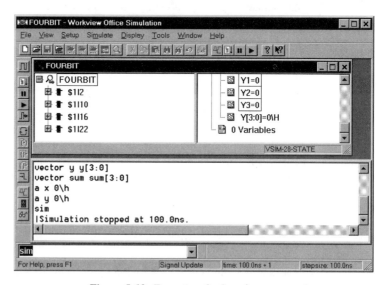

Figure 5-12: *Executing the first* **sim** *command*

1. Click on the *ViewDraw* window to view the back-annotation of the simulation results (notice that the Simulation Time in the bottom right corner has changed to 100 ns).

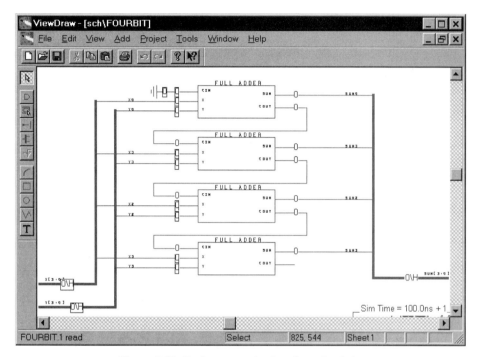

Figure 5-13: Back-annotated values from simulation

Notice that the SUM bus displays a value of zero (0), which is what we would expect from adding zero (0) to zero (0). If we did not get a correct result, we could start debugging our design. If necessary we can investigate whether the full adder components are functioning correctly. We can verify this by selecting one of the full adders and checking the underlying schematic. For example:

2. Select the top full adder by clicking on the full adder symbol.

3. Select the pop-up menu by pressing the *right* mouse button.

4. Select **Schematic**.

The schematic of the full adder is displayed along with the back-annotated simulation values for this particular instance of the full adder. Zoom in on the circuit if the values are not visible. Your schematic should look similar to Figure 5-14.

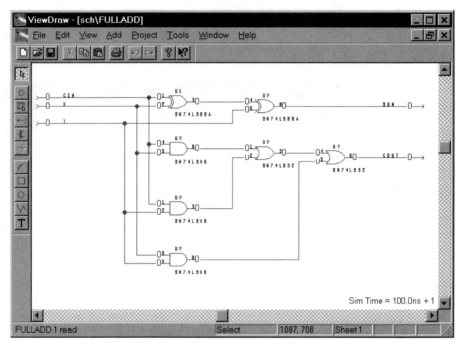

Figure 5-14: Displaying one of the back-annotated full-adder components

NOTE: The values in the underlying schematic for the full-adder component are also being simulated, and the values are back-annotated throughout the hierarchy of the design.

5.   Select **File ⇒ Close** to close the full adder schematic and return to the four-bit top-level schematic.

6.   We can now apply different input values to the X and Y buses to continue the testing of the correct operation of our design. For example, let us verify that our adder correctly adds together the values two (2) and three (3) to produce a sum of five (5). Using the assign command, assign the hexadecimal value 2 to the *x* bus and the value 3 to the *y* bus.

```
a x 2\h
a y 3\h
sim
```

7.   Click on the *ViewDraw* window to view the back annotation of the simulation results (notice that the Simulation Time has now changed to 200 ns).

Your SUM bus should display the value of 5\H as shown in Figure 5-15. If you have a different value, there is probably an error in your circuit. By looking at the fourbit and full adder schematics in *ViewDraw* (remember to press the right mouse button to activate the pop-up menu) with the back-annotated simulation values you should be able to identify any error.

> **NOTE:** If you find any errors in your full adder or fourbit design, you will need to correct the schematic, resave and check it, and create a new .VSM file (using the **Tools** ⇒ **Create Digital Netlist** command), before you can resimulate your design.

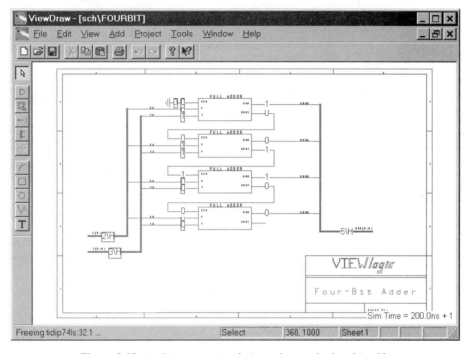

Figure 5-15: *Applying more simulation values to the four-bit adder*

Displaying Simulation Results with ViewTrace

Procedure

1. Click on the *ViewTrace* window.

2. Select the **View** ⇒ **Properties** menu command.

3. Verify the Display Values box is checked.

4. Click on **OK**.

   Your *ViewTrace* window should look similar to Figure 5-16.

5. If the complete duration of the simulation (0 to 200 ns) is not displayed, select **View** ⇒ **Full** or press **F4** for full view.

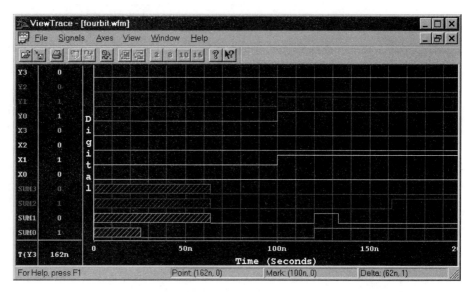

Figure 5-16: ViewTrace *window*

Notice how the X and Y signals changed from the values 0000 and 0000 to 0010 and 0011, respectively, at 100 ns. At approximately 120 ns the signal SUM0 changes and at approximately 160 ns the SUM2 signal changes, giving an output value of 0101 binary or 5. (Also notice the temporary change in the signal SUM1.) These delays are due to the signals propagating through the components in the full adder circuits.

Analyzing Waveforms

In this section, you learn how to do the following:

- Create a bus from selected signals in *ViewTrace*

- Check the time between events

- Zoom in on areas of the waveform

- Save waveform results

Creating a Bus Signal

1. Click on the *ViewTrace* window.

2. Hold down the CTRL key and click on the Y3, Y2, Y1, and Y0 signal names in the left column, which is called the *Names* region, as shown in Figure 5-17.

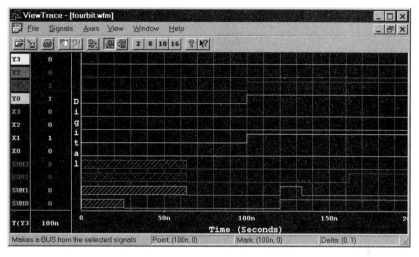

Figure 5-17: Selecting the Y signals

3. Select the *Bus* button on the toolbar. The location of this button is shown in Figure 5-17.

4. The *Signal Operations* dialog box opens:

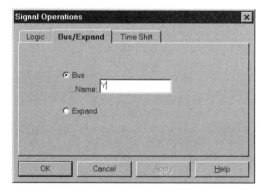

Figure 5-18: Creating a new X bus

5. Click on **Bus** and enter **Y** in the *Name* field.

6. Click on **OK.**

    A new signal called Y appears in the *ViewTrace* window, showing the binary value of the four Y signals.

7. Using similar operations create an **X** bus and a **SUM** bus from the X and SUM individual signals. To deselect the original Y signals just click on a new signal without holding down the **CTRL** key (see the following sidebar, "Selecting and Deselecting Signals," for additional information).

Once you have created the three new signals, the *ViewTrace* window should look similar to Figure 5-19.

Selecting and Deselecting Signals

Before you can perform most signal operations, you must first select the signal or signals you want to measure or modify.

- **Selecting a Signal**
  Click on the signal name in the *Names* region. After selecting the signal, its name will be displayed in reverse video within the *Names* region. This is an exclusive selection; selecting another signal in this way will deselect the previously selected signal.

- **Selecting Additional Signals**
  To select additional signals one at a time, just reposition the cursor over the signal or signal name you want to select, hold down the **CTRL** key and click the left mouse button.

- **Selecting a Range of Signals**
  To select a group of adjacent signals, select the first signal in the group then select the last signal in the group using the **SHIFT** and left mouse button.

- **Deselecting a Single Signal from a Group**
  Hold down the **CTRL** key and click the left mouse button.

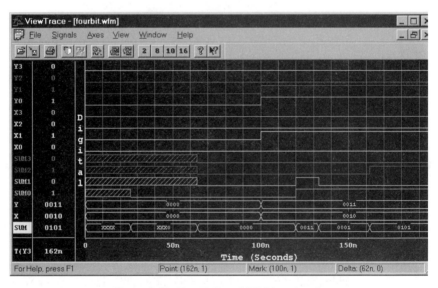

Figure 5-19: *New X ,Y, and SUM bus signals*

Checking the Time Between Events Using the Crosshair

You use the Point and Mark to measure the time difference between two signal transitions along a signal's waveform.

The Point and Mark are temporary reference points that you can position and reposition on any signals within the *Waveforms* region. You use the Point and Mark to show the X Y coordinates of the Point and Mark and the difference (delta) between the two coordinate sets. This information is displayed in the Status Bar at the bottom of the *ViewTrace* window.

For an example, we measure the delay from changing the X and Y inputs (which occurred at 100 ns) to a change in the SUM outputs.

1. Point the cursor at the rising edge of the X1 signal and click.

    The Point appears as a crosshair (+) in the *Waveforms* region. The Point is set at the current cursor location.

    Notice the Point time displayed in the status area at the bottom of the *ViewTrace* window.

2. Point the cursor at the rising edge of the SUM2 signal and click.

    The crosshair moves to that location—Notice the new time (162 ns)

3. Double-click on the rising edge of the X1 signal.

    The Point (+) and the Mark (X) are set to the same location.

4. Single-click on the rising edge of the SUM2 signal.

    The Point moves to the new location.

    The Status region of the *ViewTrace* window displays information about the Point and Mark time and the delta time between the two events (see Figure 5-20).

    This indicates that the propagation delay through our four-bit adder for a change in the inputs to a change in the outputs is 62 ns.

> **NOTE:** By default, *ViewSim* uses typical delays for the components it is simulating. For worse case delays you need to specify DELAY MAX.

> **NOTE:** Because this design is a ripple-carry adder, applying different input values may result in different propagation delays. We investigate this further in the laboratory exercises.

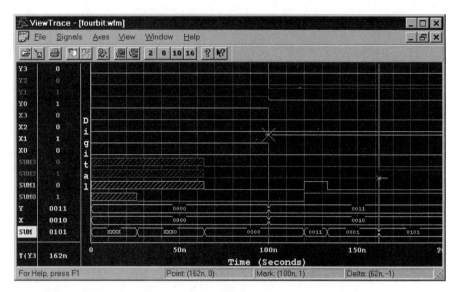

Figure 5-20: Measuring the time between events using the crosshair

Zooming in on the Timeline

You can zoom in on portions of the display using the following keys:

- **F4**—full view. Allows you to see the complete waveform traces for the current simulation period.

- **F6**—pan. Allows you to pan across the waveform traces. Point the cursor in the waveform trace area and press F6.

- **F7**—zoom in. Allows you to zoom in to display more detail.

- **F8**—zoom out.

- **F9**—select zoom range. Click on **F9**. The word "Zooming" appears in the lower left corner of the *ViewTrace* window. Drag the mouse pointer in the waveform region over the area of the waveform traces that you want to zoom in.

  You can also click and drag in the waveform time region at the bottom of the *ViewTrace* window to zoom in on part of the waveform.

Moving Signals

You can rearrange the order of the waveform signals by simply clicking on a signal name in the name region and dragging the name to a different location. The new location will be indicated by a horizontal line. Release the mouse when you reach the required location.

Changing the Width of the Name and Value Columns

You can change the width of the *Values* and *Names* regions.

You may need to widen the values column if the values are not completely visible. *ViewTrace* indicates this with dashes next to the waveform name, meaning that the values are longer than the width of the column.

1. Select the **View** ⇒ **Properties** menu command.

2. Select the **Widths** page.

3. Change the width of the *Names* or *Values* field as required.

Saving Waveform Results

You can save the *ViewTrace* waveform results in a file for later review.

1. Select the **File** ⇒ **Save** or **File** ⇒ **Save As** menu command.

   For example, you can save your waveform trace results in a *Binary Array Dump* (**.BAD**) format. This is a portable binary format useful for quickly saving and restoring *ViewTrace* data.

2. Select the **File** ⇒ **Exit** command to leave *ViewTrace*.

Conclusion

The *Workview Office* Design Process

This completes the *ViewDraw*, *ViewSim*, and *ViewTrace* Tutorial for *Workview Office*. You performed the following steps in the design process:

- Created a project in the *Workview Office Project Manager*

- Added libraries to your search path in the *Library List Editor*

- Created a schematic in *ViewDraw*

- Added components and connected them with wires (nets) and buses

- Created a symbol and an underlying schematic in *ViewDraw*

- Simulated the schematic design in *ViewSim*

- Displayed back-annotated simulation values in *ViewDraw*

- Verified the design by analyzing waveforms in *ViewTrace*

What to Do Next!

At this point you have covered an introduction to the complete design process, from the schematic entry of a design to checking the design works correctly through simulation.

It is now recommended that you briefly review the next chapter, which shows some additional *ViewDraw* and *ViewSim* commands that you will find useful when you complete the laboratory exercises or your own work.

You should also review Chapter 7, "Using *Workview Office* Help," to learn how to use the comprehensive on-line help available in *Workview Office*.

Good luck with your designs!

•

6 Additional *ViewDraw* and *ViewSim* Commands

This chapter provides an overview of some additional schematic entry and simulation commands that you may find useful when you are designing and simulating the state machines of the random number generator, traffic light controllers, or the sequential combination lock.

> **NOTE:** Remember to use the comprehensive on-line help facility if you forget how a command operates or how it is used. The help facility for each of the tools can be invoked by selecting the help icon from the toolbar or **Help** ⇒ **Contents** menu command or by pressing **F1**. The help facility is described in more detail in the following chapter, "Using *Workview Office* Help."

Additional Schematic Entry Commands

This section provides an overview of some additional schematic entry commands that you may find useful when you are creating your designs.

Changing the Sense of a Signal

It is sometimes necessary to change the *sense* of a signal to indicate that it is an *active-low* signal. This means the signal causes an action to occur when it is a logic '0' instead of a logic '1'.

For example, we need to label the net connected to the PRE input on the 74LS74A in the schematic shown in Figure 6-1. The signal has already been selected (by clicking on it). Notice the *$number* in the bottom left corner of the screen. This *$number* is the system identifier for that particular net. All signals and components have a system identifier—Try clicking on signals or components in one of your schematics.

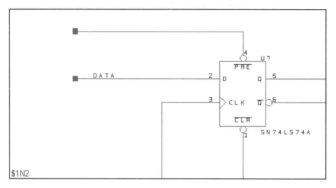

Figure 6-1: *Selecting the wire connected to the PRE pin of the 74LS74A component*

We will name the selected net PRESET, since it is connected to the PRE input on the D-type flip-flop. Notice the PRE input is active-low, indicated by the small bubble on the input and by the bar across the name PRE.

Nets are labeled by double-clicking on them. A *Net Properties* dialog box opens as shown in Figure 6-2. The label for the net has been entered as PRESET. To change the sense of the signal all you have to do is click on the *Inverted* check box, as shown in Figure 6-2.

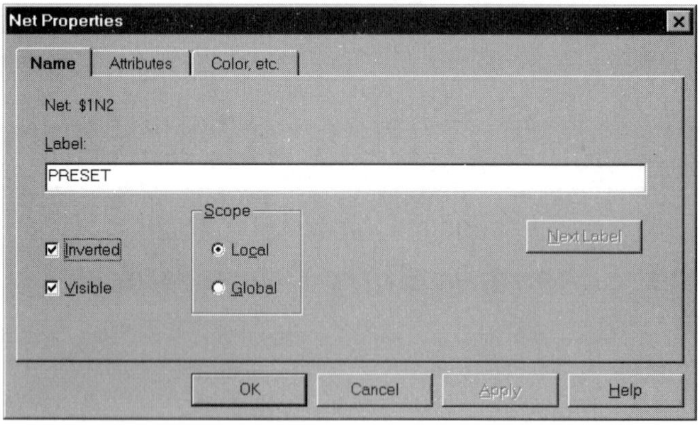

Figure 6-2: *The Net Properties dialog box*

After clicking on **OK**, the schematic is updated with the new label, as shown in Figure 6-3. (The label was moved to the right from its default position to improve the appearance of the schematic: to move a label select it and drag it.)

Notice that the inverted signal label is represented by the horizontal bar across the name. Also notice that in the bottom left corner of the screen the new name of the net is shown (~PRESET) along with the original system identifier. The tilde symbol (~) has the same meaning as the horizontal bar across the name. The tilde symbol is used when the schematic is completed and a net list is generated. Since the netlist is simply a text file, the tilde symbol is used to indicate which nets have inverted sense.

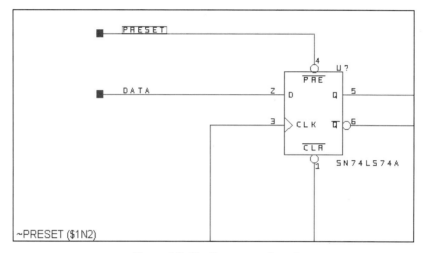

Figure 6-3: *The Preset net selected*

Assigning Attributes

Each component in a schematic has a set of default property values, or attributes. These attributes describe certain features of the component. Some attributes are used to identify physical characteristics about the component, such as the package size and layout, that will be used by a printed circuit board layout tool to generate the board. Other attributes will describe timing information, such as propagation delays that could be used by a digital simulator. You can call up all of the attributes for a given component and modify them using a dialog box—a graphical box containing a table of attributes.

We will change the REFDES or reference designator attribute on a schematic. The reference designator provides an identifier for each physical device on a schematic. Every board-level symbol will need a reference designator to enable easy packaging into physical devices and for third-party use by Printed Circuit Board (PCB) tools.

This procedure uses the full-adder design from the Schematic Entry tutorial in Chapter 4. You can, however, apply the procedure to any component on any of your designs.

1. Select the 74LS86A component in the top left corner of your schematic.

2. Select the **View > Selected** command to zoom in on the 74LS86A.

    Figure 6-4 shows the zoom (the U? is the default REFDES value):

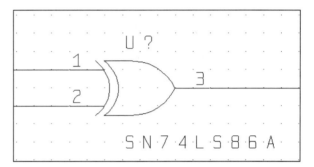

Figure 6-4: Selecting the 74LS86A component

3. Double-click on the 74LS86A component.

    The *Component Properties* dialog box is displayed:

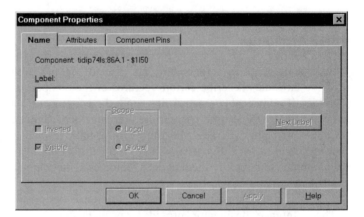

Figure 6-5: The Component Properties dialog box

4. Select the *Attributes* page:

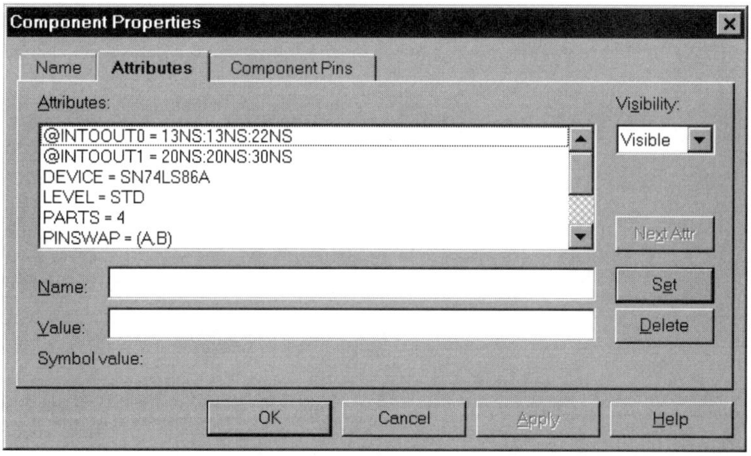

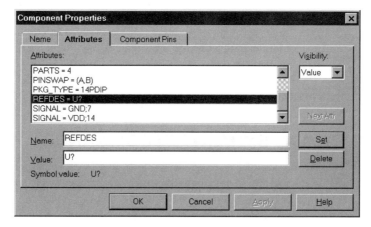

Figure 6-6: *Displaying the component attributes*

The selected component's existing attributes are listed. You can add new attributes, edit existing ones, and change default component values.

Scroll down the attribute list and select the REFDES attribute, as shown in the following figure:

Figure 6-7: *Selecting the REFDES attribute*

5. In the *Value* field, change the U? to **U1**.

6. Click on **OK** or press **Return** to return to the schematic. When you click on **OK**, the dialog box disappears and the attribute component values appear on the schematic. The REFDES attribute has now been updated to U1, as shown in Figure 6-8.

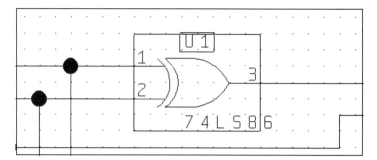

Figure 6-8: After changing the REFDES attribute

> **NOTE:** An alternative way to change an attribute like the REFDES attribute is to double-click on the attribute to display the *Attribute Properties* dialog box.

Moving an Attribute

It is sometimes necessary to move an attribute to a different location on the schematic to improve its appearance.

> **HINT:** If you cannot see the REFDES attribute, pan (**F6**) or zoom (**F7**) until you do.

1. Point the cursor at the REFDES attribute **U1** and click and hold down the left mouse button to move the attribute to a different location, then place it by releasing the left mouse button.

Click in an empty area of the schematic to deselect it.

ADDITIONAL SIMULATION COMMANDS

This section provides an overview of some additional simulation commands that you may find useful when you are simulating the state machines of the random number generator, the traffic light controllers, or the sequential combination lock.

The commands covered are

- **clock**
- **stepsize**
- **cycle**
- **pattern**
- **watch**
- **display**

Setting Up a Clock (CLOCK, STEPSIZE, and CYCLE commands)

ViewSim provides a convenient way to set up clock signals:

1. Select the **Setup** ⇒ **Stimulus** command.

    The *Setup Stimulus* dialog box displays.

2. Select the *Clock/Pattern* page.

3. In the *Signal* or *Vector Name* field enter **clk** (or the name of the signal you are using for your clock).

4. Select the *clock* option.

5. In the *Pattern* field enter the numbers **0** and **1**, separated by a space, then press **Return**.

    Click on **Add**.

    The *Setup Stimulus* dialog box should look similar to Figure 6-9.

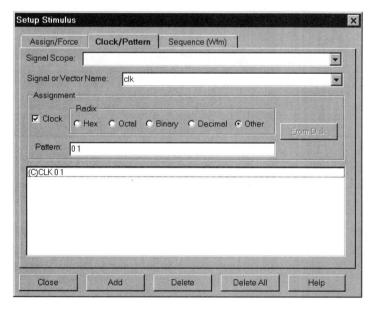

Figure 6-9: *Creating a clock signal*

The keyboard command equivalent displays in the *ViewSim* window:

clock clk 0 1

This command produces a clock signal with equal low and high periods (i.e., a clock with a 50% duty cycle). To produce a non-square wave clock simply enter a different transition sequence. For example, the keyboard command

clock clk2 0 1 1

will produce a clock wave with a 66% duty cycle, i.e., the high period being twice as long as the low period.

The duration of each of the transitions corresponds to the current value of the parameter *stepsize*. The default value of stepsize is 100 ns and it can be changed by using the **Setup ⇒ Simulator Settings** menu command (or the **stepsize** keyboard command). For example, we can change the stepsize to 50 ns using the command:

stepsize 500   I (remember that default time units are 0.1 ns, so
                      I stepsize = 50 ns)

With this stepsize the following *ViewTrace* diagram shows the waveforms for the two previously defined clocks:

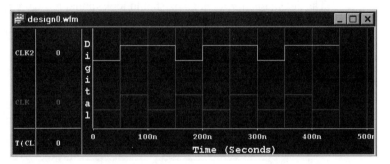

Figure 6-10: Waveforms showing two clocks

The frequency of a clock can be altered by simply changing the current value of stepsize.

Once a clock signal or signals have been created then the **cycle** command can be used to run the simulation for one full clock cycle or a specified number of cycles. The **Simulate ⇒ Run** menu command or the **cycle** command from the vertical toolbar (or the **cycle** keyboard command) can be used. For example:

         cycle 3   I simulate for three clock cycles
         cycle   I simulate for one clock cycle

If you have multiple clock signals the simulation time will be determined by the duration of the longer clock. For example, in the *ViewTrace* diagram shown in Figure 6-10 the waveforms were simulated with the command

cycle 3

Setting Up an Input Pattern (PATTERN command)

To set up an arbitrary waveform on a node or vector the *ViewSim* **pattern** command can be used.

1. Select the **Setup** ⇒ **Stimulus** command.

    The *Setup Stimulus* dialog box displays.

2. Select the *Clock/Pattern* page.

3. In the *Signal or Vector Name* field enter the name of the signal you want to assign the pattern to (for example, x).

4. In the *Pattern* field, enter the numbers **0 1 1 0 1**, separated by spaces, then press **Return**.

    The keyboard command equivalent displays in the *ViewSim* window:

    pattern x 0 1 1 0 1

    The **Simulate** ⇒ **Run** menu command (or **run** keyboard command) will apply each pattern and simulate for the current stepsize time units, until the longest pattern has been exhausted.

    For example, the following commands:

    ```
    stepsize 100ns
    vector x x[1:0]                  I create a vector with two bits
    pattern x 00 01 10 11 11 10 01 00 I apply pattern to vector
    run                              I simulate for the duration of the pattern
    ```

will produce the signals as shown in the following *ViewTrace* window:

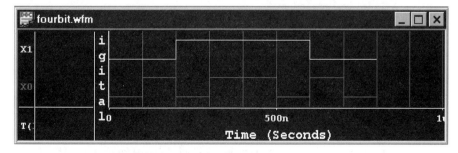

Figure 6-11: *Example waveforms generated using the Pattern command*

Watch

This command places nodes or vectors on an internal list that is accessed by the **display** and **print** commands. Anytime a signal on the watchlist is changed during a simulation run, the time when the signal changed and its new value are displayed.

Examples:

watch sum x1 x2 | places vector sum and nodes x1 and x2 on
                           | the watchlist

Display

This command lists the current value of nodes and vectors placed on the watchlist by the watch command. It may also be used to display the current value of any other node or vector.

Examples:

display | displays current values of nodes and vectors on
                      | the watch list
display x1 x2 | displays watchlist and also nodes x1 and x2

Command Files

Instead of entering commands interactively through the *ViewSim* command line, it is possible to execute a sequence of commands from a command file. Usually, this will be much faster and easier than interactively typing commands.

Figure 6-12 shows the contents of an example command file that you could use to simulate a design (I used this command file to simulate my Carry Look-Ahead circuit from Laboratory Exercise 4). The file can be created with any text editor (such as Notepad) and it should be given the extension .cmd. Save the command file in the c:\LABS directory or your current project directory.

```
| Typical delays in use.
| All delays scaled by 1.
| Total of 20 digital modules were processed.
wave LOOKAHD.wfm x0 x1 x2 x3 y0 y1 y2 y3 sum0 sum1 sum2 sum3
vector x x[3:0]
vector y y[3:0]
vector sum sum[3:0]
watch x y sum
pattern x 0 0010 0101 1100 1111 0001 0000 0001
pattern y 0 0011 0001 0110 1011 1111 1111 1111
run
```

Figure 6-12: Example of command file (lookahd.cmd) for the Look-Ahead Carry circuit

You can execute command files in *ViewSim* with the **File ⇒ Run Command File** command from the *ViewSim* **File** pull-down menu.

The first three lines that start with a vertical bar are simply comment lines and will be ignored by the simulator.

The **wave** command will generate a waveform to link to *ViewTrace* and automatically start *ViewTrace* running. The name of the waveform is specified as LOOKAHD.WFM and the signals to be displayed in *ViewTrace* are x0, x1, x2, etc.

The next three commands are **vector** commands and will create vectors for the X, Y and SUM buses, (creating vectors makes it easier to modify and manipulate signals).

We next issue a **watch** command for the x, y, and sum vectors. Anytime these signals change value, the time of the change will be displayed.

Notice the **pattern** commands to apply the different values to the X and Y buses in sequence.

Finally, the **run** command will run the simulation for the length of time determined by the number of different patterns in the pattern command. Each pattern will be applied for the default simulation stepsize of 100 ns, resulting in a total simulation duration of 800 ns.

> **NOTE:** *ViewSim* automatically saves all commands that you enter in the VIEWSIM.LOG file in the current project directory. This log file is simply a text file with all your interactive commands. It also contains all the system responses placed as comments after a vertical bar. You can edit this file to add or modify any commands and then rename this file to *filename.CMD* and execute it as a command file in a new simulation session.

"Circuit Not Yet Stable"

If after simulating a circuit for a period of time a "circuit not yet stable" message appears, the signals are still propagating through the circuit and you need to run the simulation for a longer period of time. This may be a perfectly valid condition and is just a result of the number of gates in your circuit.

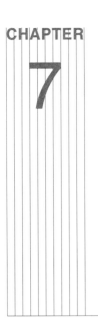

CHAPTER

7

Using *Workview Office* Help

The *Workview Office* product has comprehensive on-line help available from any of the three main tools: *ViewDraw*, *ViewSim*, and *ViewTrace*.

For example, in *ViewDraw*, the help facility can be invoked by selecting the help icon from the toolbar (as shown in Figure 7-1) or by the **Help** menu command or by pressing **F1**.

Figure 7-1: The ViewDraw help icons on the toolbar

For example, pressing **F1** (with no schematics open) displays the general *ViewDraw* help menu:

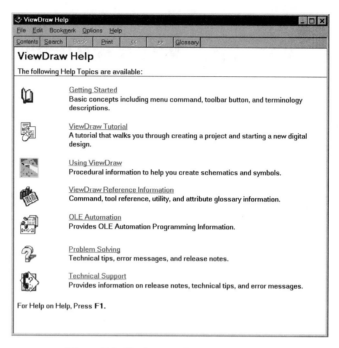

Figure 7-2: Workview Office help dialog box

There are sections on *Getting Started*, a *ViewDraw Tutorial*, reference information, etc.

Selecting the **Help ⇒ ViewDraw Help Topics** menu command from *ViewDraw* will display the *Help Topics: ViewDraw Help* window, where the same help information can be accessed in book format.

Figure 7-3: Help Topics: ViewDraw Help

You can also use the **Index** or **Find** tabs from this window to search for information on a particular command or feature.

You can use the *ViewDraw* help query command by clicking on the *ViewDraw* help icon in the toolbar. You can then click the question mark that appears on an icon or area of the window to find out more information about that object. For example, positioning and clicking the question mark on the *Add Component* icon on the left toolbar displays the following information about the component command:

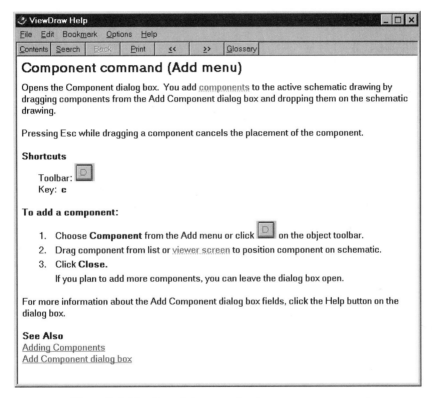

Figure 7-4: ViewDraw help screen for the component command

HINT: Notice the equivalent keyboard command (Key: **c**) for the *Add Component* command. Many of the *ViewDraw* commands have equivalent keyboard shortcuts to save using the mouse pointer. You may want to experiment with these commands.

The *ViewSim* and *ViewTrace* tools also have similar on-line help commands.

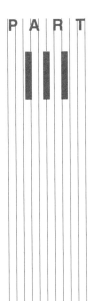

PART

III

LABORATORY EXERCISES

LABORATORY EXERCISES

Laboratory Exercise 1:
Introduction to Digital Logic Devices and Their Operation

Laboratory Exercise 2:
Design of a Four-Bit Ripple-Carry Adder

Laboratory Exercise 3:
Design of a Four-Bit Arithmetic Unit

Laboratory Exercise 4:
Design of a Look-Ahead Carry Adder

Laboratory Exercise 5:
Design of a Four-Bit ALU

Laboratory Exercise 6:
Design of a Random Number Generator

Laboratory Exercise 7:
Design of a Simple Traffic Controller

Laboratory Exercise 8:
Design of a Traffic Controller with Left Turn Signal

Laboratory Exercise 9:
Design of a Sequential Combination Lock

Laboratory Exercise 10:
Design of a Vending Machine

OVERVIEW

In each of the following laboratory exercises there is a section briefly covering the theoretical issues involved in the designs. This information is not meant to be a complete or thorough examination of the topics. Instead, the aim is to remind the designer of the important issues that are being covered in the exercises. It is recommended that a conventional textbook be used for reference during the design process.

The following are text books used by the author that may be useful.

Digital Design Principles and Practices by John F. Wakerley, second edition, published by Prentice Hall.

Digital Design by M. Morris Mano, second edition, published by Prentice Hall.

Contemporary Logic Design by Randy H. Katz, published by Benjamin Cummings.

8 Laboratory Exercise 1: Introduction to Digital Logic Devices and Their Operation

Description

This laboratory exercise introduces you to simple digital logic devices and their operation. It also introduces you to the process of building digital logic circuits using a digital design kit. Such a kit has a number of breadboards in which the integrated circuits can be placed and connected together by wires. In addition, there are indicator lights, usually light emitting diodes (LEDs), to show the state of the output signals, and switches for the inputs.

Procedure

Investigation of Simple Logic Gates

The following table lists a number of TTL logic devices and shows the pin locations of the inputs and outputs for the devices. All of the integrated circuits have four gates per package, except the 74LS04, which has six gates. Each of the four (or six) gates are identical in each package.

For all of the integrated circuits, GND is pin 7 and VCC (5V) is pin 14.

	Input Pin Numbers						Output Pin Numbers					
	Gate 1	Gate 2	Gate 3	Gate 4	Gate 5	Gate 6	Gate 1	Gate 2	Gate 3	Gate 4	Gate 5	Gate 6
74LS00	1,2	4,5	9,10	12,13			3	6	8	11		
74LS04	1	3	5	9	11	13	2	4	6	8	10	12
74LS08	1,2	4,5	9,10	12,13			3	6	8	11		
74LS32	1,2	4,5	9,10	12,13			3	6	8	11		
74LS86	1,2	4,5	9,10	12,13			3	6	8	11		

The lab instructor will provide you with one of each of the integrated circuits listed in the previous table.

For each of the integrated circuits complete the following truth tables showing the operation of one of the gates in each integrated circuit. You can do this by connecting the inputs to switches and the outputs to LEDs on your development board. Also describe in words the operation of the gates. (Notice that the operation of each of the four or six gates in each integrated circuit is identical. You can verify this for yourself by selecting a different gate.)

74LS00

Input 1	Input 2	Output

Describe the operation of this device:

74LS04

Input	Output

Describe the operation of this device:

74LS08

Input 1	Input 2	Output

Describe the operation of this device:

74LS32

Input 1	Input 2	Output

Describe the operation of this device:

74LS86			Describe the operation of this device:
Input 1	Input 2	Output	

Design of a Simple Controller

With information from the previous section, design a circuit with multiple gates to implement the following:

You are to design a logic circuit that will be part of a control system. It has four inputs that we will call A, B, C, and D and one output that we will call E. The output must be ON when input A is ON and input D is OFF and either B OR C (or both) is ON—For all other input conditions the output must be OFF. Using only the integrated circuits that you have worked with in the previous section, design and construct a logic circuit to carry out this control operation. Use four switches for the inputs and one LED for the output. Demonstrate to the lab instructor that your circuit operates correctly and complete the following table:

Input A	Input B	Input C	Input D	Output E
OFF	OFF	OFF	OFF	
ON	OFF	OFF	OFF	
OFF	ON	OFF	OFF	
ON	ON	OFF	OFF	
OFF	OFF	ON	OFF	
ON	OFF	ON	OFF	
OFF	ON	ON	OFF	
ON	ON	ON	OFF	
OFF	OFF	OFF	ON	
ON	OFF	OFF	ON	
OFF	ON	OFF	ON	
ON	ON	OFF	ON	
OFF	OFF	ON	ON	
ON	OFF	ON	ON	
OFF	ON	ON	ON	
ON	ON	ON	ON	

Draw your circuit diagram. Include IC and pin numbers. You do not have to include power connections.

Simple Sequential Device

The devices you have used in the previous sections are used to design circuits whose outputs depend only on the current conditions of the input signals—these circuits are known as *combinational* circuits.

We now briefly investigate the operation of a flip-flop, which is an element that can be used in the design of *sequential* circuits—circuits whose outputs depend not only on current input conditions but also on previous input conditions. These types of circuits are said to have memory; e.g., they remember previous input conditions.

A 74LS74 integrated circuit has two D-type flip-flops within its 14-pin package. We will connect one of these flip-flops to switches and LEDs as follows.

Connect pin 2 (this pin is connected to the D input of the device) to one switch.
Connect pin 3 (this pin is connected to the CLK input of the device) to a second switch.
Connect pin 5 (this pin is connected to the Q output of the device) to one LED.
Connect pin 6 (this pin is connected to another output of the device) to a second LED.

Try different switch input combinations and monitor the state of the two LEDs for each of the combinations.

Describe in table form and in words the operation of this type of logic device.

Laboratory Exercise 2: Design of a Four-Bit Ripple Carry Adder

Description

The goal of this exercise is to design a simple combinational circuit that can add two four-bit binary numbers together. This circuit will be used in a later exercise (with some modifications) as part of a more complex arithmetic logic unit (ALU) that is representative of ALUs found in modern microprocessors.

This exercise will also introduce you to the Viewlogic digital design and analysis tools. You will learn how to use the Viewlogic *ViewDraw* schematic entry software to draw your design. Once the design is completed you will use the simulation and waveform analysis tools, *ViewSim* and *ViewTrace*, to demonstrate your circuit functions correctly.

Theoretical Background

To get you started thinking about binary addition, add together the following decimal numbers (without using your calculator!):

$$\begin{array}{r} 1234 \\ + \ 5678 \\ \hline \end{array}$$

We assume you started with the least significant digits (the units, or 4 and 8 in this example). Adding 4 to 8 produces twelve or two and a ten. This carry of ten is added to the next column (the tens) giving us 3 and 7 and 1 (the carry from the units column) which is eleven or 1 and a carry of 1. We continue on in this manner for the remaining columns to obtain the result of 6912.

In binary addition the process is exactly the same but instead of working with decimal numbers that can have values from 0 to 9, we are working with binary numbers which can only have values of 0 and 1. This means that if we add 1 and 1 together (which would produce a value of 2 in the decimal numbering system) we obtain a sum of 0 and a carry of 1. All the possible combinations for adding two 1-bit binary numbers together can be shown in a truth table as follows (where the two numbers are shown represented by X and Y):

X	Y	CARRY	SUM
0	0	0	0
0	1	0	1
1	0	0	1
1	1	1	0

Table 9-1: *Truth table of a half adder*

This type of operation which can add together two 1-bit numbers and produce a sum and carry output is called a half adder. A block diagram of a half adder is shown in Figure 9-1.

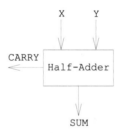

Figure 9-1: *Block diagram of a half adder*

A half adder circuit can be designed with a few simple gates. The carry output (which needs to produce a '1' output when both of the inputs are '1') can be obtained by 'anding' together the two inputs. The sum output (which needs to produce a '1' output when one input, but not both of the inputs, is a '1') can be obtained with an EXCLUSIVE-OR gate. A schematic (circuit diagram) of a half adder is shown in Figure 9-2.

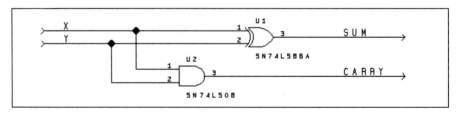

Figure 9-2: *Circuit diagram of a half adder*

This type of adder is suitable for adding together the least significant bits of a binary number but it does not have the capability to be used for the remaining bits of a number where we need to not only add together two 1-bit numbers but we also need to be able to add any carry-in from the less significant bit positions.

A circuit known as a full adder has the required functionality. It can take two 1-bit binary numbers, say X and Y, and a possible carry-in, CI, and sum the 3 inputs together. The result can take on values from 0 to 3, and thus requires 2 bits, CO and SUM, for representation. CO is the most significant bit, and designates the carry-out, and SUM is the sum bit for a given column of addition. The truth table of a full adder is shown in Table 9-2.

X	Y	CI	CO	SUM
0	0	0	0	0
0	0	1	0	1
0	1	0	0	1
0	1	1	1	0
1	0	0	0	1
1	0	1	1	0
1	1	0	1	0
1	1	1	1	1

Table 9-2: Truth table of a full adder

A block diagram of a full adder is shown in Figure 9-3.

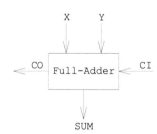

Figure 9-3: Block diagram of a full adder

The circuit implementation of a full adder is slightly more difficult than the half adder. The carry out (which has to produce a '1' output when at least two of the inputs are '1') can be obtained by using three two-input AND gates and a three-input OR gate. The sum output (which has to produce a 1 output when one or three inputs—i.e. an odd number of 1's—are equal to '1') can be obtained by using two 2-input EXCLUSIVE-OR gates. The circuit implementation of a full adder is shown in Figure 9-4. It can be seen that two 2-input OR gates are used instead of a single 3-input OR gate because 3-input OR gates are not available in 74LS technology integrated circuits.

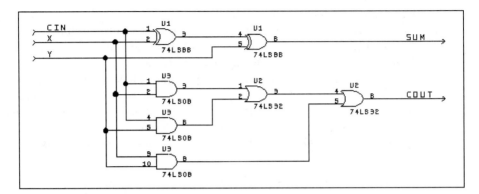

Figure 9-4: Circuit diagram of a full adder

There are a number of different circuit implementations that would achieve the same logic result as represented by the truth table in Table 9-2. One practical note—you may have wondered why we did not use a three input EXCLUSIVE-OR gate instead of the two 2-input EXCLUSIVE-OR gates. Well, for the same reason we had to use two 2-input OR gates earlier; IC manufacturers do not make three-input EXCLUSIVE-OR gates!

    Once we have completed the design of a half adder and a full adder we can connect as many of these together as we require to make a multi-bit adder. For example a four-bit adder may be constructed from one half adder and three full adders as shown in Figure 9-5.

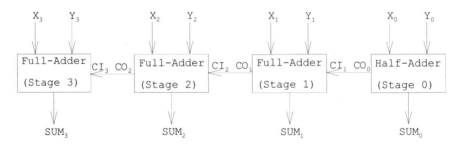

Figure 9-5: Block diagram of a four-bit adder

It can be seen that the addition of the least significant bits in the first stage only requires a half adder since there is no carry-in to this stage. If a full adder was used also for the least significant stage then the carry-in input would be connected to logic '0'.

Negative Numbers—Two's Complement Representation

In the above discussion of adder circuits we assumed we were working with positive numbers. This means that with a four-bit adder we can work with binary numbers from 0000 (zero) to 1111 (fifteen in decimal). However it would be useful if we could also add together negative numbers.

We can represent negative binary numbers in a variety of different ways but the best representation from a circuit point of view is known as two's complement. In two's complement representation the sign, positive or negative, can be determined by the most significant bit (MSB). If the MSB is zero then the number is positive, and if the MSB is 1 then the number is negative. A positive binary number in two's complement representation is identical to an unsigned binary number but a negative number is obtained by first complementing the bits of the positive number and then adding one. Any carry out of the MSB is ignored.

Let us look at some four-bit examples, showing how we obtain the two's complement of binary numbers:

Converting 2 to -2:

	decimal 2	0010	
		1101	complement bits
		1	add 1
	decimal -2	1110	

Converting 5 to -5:

	decimal 5	0101	
		1010	complement bits
		1	add 1
	decimal -5	1011	

Converting 6 to -6:

	decimal 6	0110	
		1001	complement bits
		1	add 1
	decimal -6	1010	

Converting -4 to 4:

	decimal -4	1100	
		0011	complement bits
		1	add 1
	decimal 4	0100	

The complete range of binary numbers available using four-bits is shown in Table 9-3:

Decimal Number	Binary Number	Two's Complement Representation
-8		1000
-7		1001
-6		1010
-5		1011
-4		1100
-3		1101
-2		1110
-1		1111
0	0000	0000
1	0001	0001
2	0010	0010
3	0011	0011
4	0100	0100
5	0101	0101
6	0110	0110
7	0111	0111

Table 9-3: Four-bit two's complement numbers

Notice that the two's complement number representation is not symmetrical; there is always one more negative than positive number.

Also, notice that we can obtain the next higher number in the series by just adding one to the previous number (ignoring any carries that are generated in the process). And, because adding a number is just the same as adding one repetitively the required number of times, we can use the same adder circuit we previously discussed for both unsigned binary numbers and also two's complement numbers.

Procedure

Using the schematic entry tool called *ViewDraw,* draw a schematic of a four-bit binary adder. It is recommended that you use the tutorial introduction to the *ViewDraw* section (Chapter 4) if you are not already familiar with using *ViewDraw*. This chapter guides you step by step through the process of designing a four-bit adder, while at the same time showing you how to use the commands and features of *ViewDraw*.

The four-bit binary adder will have two four-bit input numbers and will produce a sum output also four bits in length. The two 4-bit inputs are **X[3:0]** and **Y[3:0]** and the sum is **SUM[3:0]**. These signals are 4-bit buses and the bracketed numbers indicate the individual bits in the bus, for example, SUM3, SUM2, SUM1, and SUM0, where SUM3 is the MSB, and SUM0 the LSB of the sum.

Instead of drawing a large single-sheet design (which would be cumbersome to draw and difficult to understand) we want to design a hierarchical circuit. This means that we will draw a one-bit full adder circuit, make a symbol to represent this circuit, and then replicate it four times for our complete four-bit adder design. Your final schematic should look similar to Figure 9-6. The X and Y inputs can be seen on the left of the circuit, while the SUM outputs are on the right.

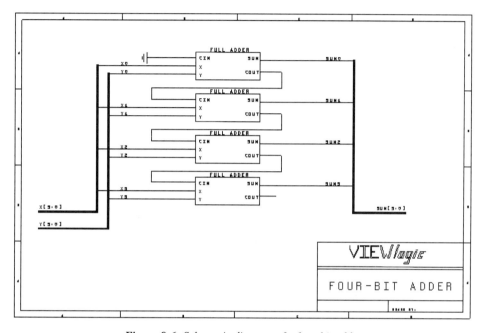

Figure 9-6: Schematic diagram of a four-bit adder

> **NOTE:** You are allowed to use only SSI devices for your design—for example, 2-, 3- or 4-input NAND, OR, XOR gates. No MSI devices (binary adders, for example) are allowed—for obvious reasons!

> **NOTE:** Only use integrated circuits from the *74LS* library (alias **tidip74ls**)—Do not select gates from the *Builtin* library because these do not represent physical devices and so do not have the necessary attributes for timing analysis.

Analyzing Your Design

Once you have finished drawing your circuit we want to make sure that it works correctly. You could do this by actually building the circuits using a collection of 74LS (or other family of devices) integrated circuits.

However, it is now standard practice, and a lot more efficient, to use the computer and software analysis tools to simulate and test the circuit. It is a lot less time-consuming and less expensive to correct errors found during simulation before the system is built.

We will use Viewlogic's interactive digital simulation tool, *ViewSim*, to test the four-bit adder circuit and the waveform generation and analysis tool, *ViewTrace*, to graphically show the operation of your circuit.

ViewSim reads in a description of your circuit and allows you to apply different input stimuli to the design. *ViewSim* then works out what all the logic values at every single node in your circuit should be for the set of inputs currently being applied. As well as the value of the logic signals (1 or 0), *ViewSim* also calculates at what time these values would occur. The timing delays of a circuit are of course dependent on the type of logic technology used. For example, a circuit made out of 74LS (low-power schottky) devices will operate slower than a circuit made out of 74S (schottky) devices. *ViewSim* reads the type of device you have selected from your schematic and from its internal libraries assigns the correct delays.

Using the *ViewSim* and *ViewTrace* tools (see the corresponding tutorial guide in Chapter 5) it is easy to confirm that the circuit is functioning as intended. In each case, apply the required input values to the X and Y buses, run the simulation of your circuit for 100 ns and then look at the SUM values produced. By applying the inputs shown in the following table, write down the sum outputs obtained to verify that your adder is working correctly:

X[3:0]		Y[3:0]		SUM[3:0]	
Decimal	Binary	Decimal	Binary	Decimal	Binary
2	0010	3	0011		
5	0101	1	0001		
-4	1100	6	0110		
-1	1111	-5	1011		
1	0001	-1	1111		

Table 9-4: Verifying correct operation of a four-bit adder

NOTE: If after simulating a circuit for a period of time a "circuit not yet stable" message appears, the signals are still propagating through the circuit and you need to run the simulation for a longer period of time.

NOTE: If during the simulation process you find that your circuit is not working correctly, you can modify your schematics and resimulate as described in the tutorial chapters.

Timing Analysis

The preceding section carried out a functional analysis of your circuit to verify that the circuit did indeed perform the addition operation correctly. This type of analysis is usually insufficient on its own. It is usually necessary to verify also that the circuit will operate fast enough for the intended application. In this section we investigate how fast our four-bit adder can work. We see that the actual time required to add two numbers together in this type of ripple-carry adder depends on the values of the two numbers.

Use the input values from the following table to calculate how long it takes for your circuit to achieve steady state (i.e., outputs are stable and no longer changing). In each case, issue a **restart** command to start the simulation again at 0 ns. For the last combination (adding 1 to -1) apply 0000 on the X bus and 1111 on the Y bus and simulate until the circuit is stable; then change the X bus to 0001.

X[3:0]		Y[3:0]		Time for SUM[3:0] outputs to
Decimal	Binary	Decimal	Binary	become stable (ns)
2	0010	3	0011	
5	0101	1	0001	
-4	1100	6	0110	
-1	1111	-5	1011	
1	0001	-1	1111	

Table 9-5: *Timing analysis of a four-bit adder*

The last input combination given in the table is the worst-case addition for this type of ripple adder, where the carry has to propagate from the least significant stage to the most significant stage. Using *ViewSim* and *ViewTrace* obtain a timing diagram showing the time taken for each of the SUM outputs (0, 1, 2, and 3) to change to their steady state. If we had a 16-bit or 32-bit adder, instead of our simple four-bit adder, such a circuit would soon suffer from intolerable delays. In a later exercise we will see how we can design an alternative adder whose delays are not dependent on the word size of the numbers we are adding together.

Obtain an expression for the worst-case delay of a ripple adder in terms of the number of stages in the adder.

Laboratory Construction (Optional)

Once a circuit has been designed and tested through simulation we could then go ahead and build it, and, assuming there were no construction errors or faulty devices, the circuit should work correctly the first time.

In this part of the laboratory exercise we will construct part of our four-bit binary adder and show it works correctly by applying different binary numbers with switches and showing the result on LEDs (light emitting diodes).

To save construction time and to reduce wiring complexity we will construct a four-bit adder without the carry-in on the LSB and without the carry-out on the MSB.

PRELAB:

For Prelab you need a copy of a schematic showing your four-bit adder. This should be a flat design (no symbols) and with the REFDES numbers for the ICs (U?) changed to represent the actual devices used. You do not need to include a carry-in for the LSB or the carry-out circuitry for the MSB.

To change the REFDES number (U?), double-click on the U? text. The *Attribute Properties* dialog box opens. Change the REFDES attribute from U? to U1, or U2, etc., as required (see the *Assigning Attributes* section in Chapter 6 for additional information).

You can also change the slot of the component. Select the component, then select the **Edit** ⇒ **Slot** menu command (see the *Adding More Components* section in Chapter 4).

Your design should probably use about five integrated circuits. The design should easily fit onto one page. Create a new design (call it *design1.1*) and use the copy and paste commands to copy parts of your previous lab design.

LAB:

We show the adder works correctly by applying the following four-bit two's complement numbers with switches and showing the result on LEDs.

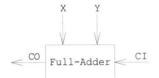

Demonstrate the correct operation of your circuit to the lab instructor.

Use a logic analyzer to obtain the propagation delay from a change on the LSB to the MSB. Apply 0000 on the X BUS and 1111 on the Y BUS, then change X to 0001. By triggering the logic analyzer when X0 changes from a logic '0' to a logic '1' you can determine how long it takes for each of the SUM outputs to change.

Compare the experimental timing results with your simulated timing values. Are they the same? Discuss the reasons for any differences.

10 Laboratory Exercise 3: Design of a Four-Bit Arithmetic Unit

Description

The goal of this exercise is to extend the four-bit adder design from Laboratory Exercise 2 to include subtraction capability. We also want to add some additional signals. After the completion of many types of arithmetic or other types of operation it is sometimes necessary to check if the result was negative or equal to zero, or if an overflow occurred. This type of testing may be required if the arithmetic unit was part of a microprocessor that was executing a program and it was necessary to branch to another part of the program depending on the result of the operation. To help with this process it is common for arithmetic circuits to provide special signals, called flags, that indicate the type of result generated. For example, we could have a zero flag that indicated if the result of the previous arithmetic operation was zero.

We add the following flags to the design of our arithmetic unit:

- **Zero flag**—Set if the arithmetic operation is zero; reset otherwise.
- **Carry flag**—Set if the carry out of the most significant stage is 1; reset otherwise.
- **Overflow flag**—Set if the result of the arithmetic operation overflowed the number range of our system (see below).

In each case we assume that *set* means that the corresponding signal will be a binary '1' and *reset* means the signal will be a binary '0'.

141

We select the type of operation we want our arithmetic unit to perform, either addition or subtraction, by having a control input signal, called *add/sub*. When this input signal is a logic '1', it indicates that we want to perform an addition operation, and when the signal is a logic '0', it indicates that we want to perform a subtraction operation.

Once we have completed the design of the arithmetic unit we prove it is functioning correctly by analyzing its operation through simulation using *ViewSim* and *ViewTrace*.

Theoretical Background

We need to review two main aspects of the design of this type of arithmetic unit, incorporating the subtraction capability and designing the appropriate flag generation circuitry.

Subtraction Capability

In the previous laboratory exercise we designed and tested a four-bit adder. We now want to modify this circuit so it can not only add binary numbers but also subtract binary numbers. However, instead of performing the subtraction operation, $X - Y$, we can achieve the same result by *adding* the two's complement of Y. The two's complement of a binary number can be obtained by complementing each of the individual bits and adding 1. These observations mean that we do not have to build a special subtraction circuit, but instead can use the previously designed multi-bit adder with minor modifications.

Before we investigate the design implications of adding a subtraction capability let us work through the steps required to convert an addition operation into a subtraction.

For example, assume we want to subtract 3 from 5. These values are shown below as decimal and binary numbers:

$$5 \qquad 0 \ 1 \ 0 \ 1$$
$$- \ 3 \qquad 0 \ 0 \ 1 \ 1$$

This operation is the same as adding 5 to -3. As mentioned above, to obtain the two's complement of a number we need to complement each of the individual bits in that number and then add 1.

$$0 \ 0 \ 1 \ 1$$
complementing individual bits $\Rightarrow$ $1 \ 1 \ 0 \ 0$
adding one $\Rightarrow$ $1 \ 1 \ 0 \ 1$

So we can rewrite the original subtraction operation as an addition, i.e.,

$$5 \qquad\qquad 0 \ 1 \ 0 \ 1$$
$$+ \quad \underline{-3} \qquad\qquad \underline{1 \ 1 \ 0 \ 1}$$
$$2 \qquad\qquad 1 \ \underline{0 \ 0 \ 1 \ 0}$$

which gives the desired result of 2, after discarding any carry generated.

Once we understand that we can subtract a number by adding the two's complement of that number, we can think about what changes will be necessary to our multi-bit adder.

As mentioned above, when we want to perform subtraction (indicated by the *add/sub* control input signal being '0'), we need to obtain the two's complement of the subtrahend (the number we are subtracting). We can then use the same adder circuit we have already designed. The two's complement of a binary number may be obtained by complementing each of the individual bits of the number and then adding one.

A block diagram of the required circuit is shown below.

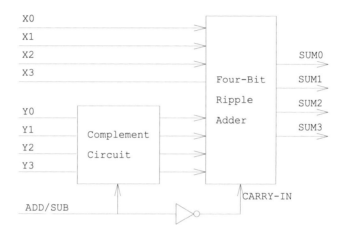

Figure 10-1: *Block diagram of an arithmetic unit*

When the *add/sub* control line is '1' then the 4-bit Y input should be passed unchanged through to the adder. In this case the circuit will perform identically to the previous design. However, when the *add/sub* line is '0', then the complement circuit should complement each of the bits prior to passing them to the adder. Also note that the extra '1' that needs to be added to the complemented input to obtain the two's complement may be incorporated by using the carry-in of the least significant stage. (If you used a half adder for the least significant stage of your four-bit adder it will have to be changed to a full adder for this circuit.)

Flag Generation Circuitry

The ZERO and CARRY flags are relatively simple to add to your design.

For the ZERO flag only one or two gates are required to produce a logic '1' output when all the sum outputs are logic '0'.

The CARRY flag is even simpler in that it may be obtained directly from the carry-out of the most significant stage.

The OVERFLOW flag is also relatively simple in its implementation, but we should briefly review why it is required. In any arithmetic circuit we design it is possible to add or subtract two numbers for which the result cannot be represented in the number range of the circuit. For example, with a four-bit binary adder, using two's complement numbers, the maximum positive number that can be accommodated is seven (0111). This means that if we were to add two positive numbers together such that the result would be greater than 7, the circuit would not give the correct answer. For example, let us add together 5 and 4:

$$
\begin{array}{r}
5 \\
+ \ \ 4 \\
\hline
9
\end{array}
\qquad
\begin{array}{r}
0 \ \ 1 \ \ 0 \ \ 1 \\
0 \ \ 1 \ \ 0 \ \ 0 \\
\hline
1 \ \ 0 \ \ 0 \ \ 1
\end{array}
$$

Unfortunately, the two's complement result of 1001 is -7, not 9. The same thing can happen with other combinations. For example, if we add -3 to -7, we get

$$
\begin{array}{r}
-3 \\
+ \ \ -7 \\
\hline
-10
\end{array}
\qquad
\begin{array}{r}
1 \ \ 1 \ \ 0 \ \ 1 \\
1 \ \ 0 \ \ 0 \ \ 1 \\
\hline
1 \ \ 0 \ \ 1 \ \ 1 \ \ 0
\end{array}
$$

The result this time is 0110 (after discarding the carry) which is 6, not -10.

These types of operations, where the answer cannot be represented by the word length of the circuit, result in an *overflow* condition. This is not a major problem if the circuit can provide some indication when the answer given is not correct. Fortunately, some simple tests can be carried out on the operands to indicate if overflow will occur. The simplest test from a circuit implementation point of view is the following:

An overflow condition exists if the sign of the operands (positive or negative, represented by the MSB) are the same and the sign of the answer (also represented by the MSB) is different than the sign of the operands. These conditions may be observed in the examples given above. To implement this test requires a few simple gates.

Procedure

Modify the four-bit adder you designed in Laboratory Exercise 2 so it can now also subtract as well as add. You are to add a control signal called add/sub to indicate whether we want the circuit to add the two four-bit numbers together (when add/sub is logic '1') or to subtract the two numbers (when add/sub is logic '0').

> **HINT:** The only modification you should have to make to your 1-bit adder circuits is the addition of an EXCLUSIVE-OR gate (to obtain the complement of the bit when subtracting). Also think about using the carry-in signal to the least significant stage (which was previously just connected to ground) to add in the extra 1 required.

Also add circuitry to your circuit to provide ZERO, CARRY, and OVERFLOW flags.

Using *ViewSim* and *ViewTrace,* demonstrate that your circuit works correctly by carrying out the following six operations:

- 2 + 3
- 1 − 5
- 4 + 4
- 2 + −3
- 6 − 6
- 2 − −3

Complete the following table:

X[3:0]		Y[3:0]		Operation	SUM[3:0]		FLAGS		
Decimal	Binary	Decimal	Binary		Decimal	Binary	Zero	Carry	Overflow
2	0010	3	0011	Add					
1	0001	5	0101	Sub					
4	0100	4	0100	Add					
2	0010	-3	1101	Add					
6	0110	6	0110	Sub					
2	0010	-3	1101	Sub					

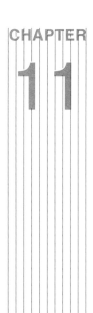

11

Laboratory Exercise 4: Design of a Look-Ahead Carry Adder

Description

In Laboratory Exercise 2 you designed a four-bit binary adder. This type of adder is called a ripple-carry adder because of the way any carries that are produced ripple through the stages from the least significant stage to the most significant stage. For example, one of the additions you carried out using the adder was to add -1 (1111 in two's complement representation) to 1. This addition is reproduced below for your convenience:

```
    0  0  0  1
+   1  1  1  1
1   0  0  0  0
```

In this particular example, each of the full-adder stages produces a sum of zero and a carry of one. This carry is then applied to the carry-in of the next stage. Although this method is simple to understand and implement it is too slow for any practical arithmetic circuit with more than a few bits. In the worst case scenario, given in the addition example above, a carry must propagate from the least significant stage of the adder to the most significant stage as shown in the block diagram below.

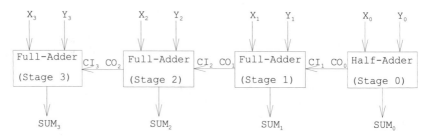

Figure 11-1: Block diagram of a ripple-carry adder

It was seen in the previous laboratory exercise that the worst-case delay was

$$T_{delay} = T_{XYCO} + (N-2) \times T_{CICO} + T_{CISUM}$$

where N is the total number of stages and T_{XYCO} is the delay from X or Y to the CO of the least significant stage, T_{CICO} is the delay for the middle stages from CI to CO, and T_{CISUM} is the delay from CI to the SUM output of the most significant stage.

Instead of using a ripple-carry scheme we investigate an alternative method of carry generation called look-ahead carry. In look-ahead carry the carry-in for each of the stages in a multi-bit adder is generated directly from the addends instead of from the carry-in of the preceding stage. In look-ahead carry circuits we usually define two signals for each full adder: a carry-generate signal, G_i (for stage i), and a carry-propagate signal, P_i. The carry-generate signal of a particular stage will *generate* a carry if the two input bits of that stage are both logic 1. The carry-propagate signal will *propagate* a carry signal if one (or both) of the input bits are 1. These conditions may be shown by referring to the truth table shown in Table 11-1.

CI_i	X_i	Y_i	Carry-Generate (G_i)	Carry-Propagate (P_i)
0	0	0	0	0
0	0	1	0	1
0	1	0	0	1
0	1	1	1	1
1	0	0	0	0
1	0	1	0	1
1	1	0	0	1
1	1	1	1	1

Table 11-1: Truth table of full adder showing carry-generate and carry-propagate signals

The logic equations for these two signals are

$$G_i = X_i . Y_i$$
$$P_i = X_i + Y_i$$

and so the carry-out of any stage of a multi-bit adder can be written in terms of these signals:

$$CO_i = G_i + P_i.CI_i$$

This may be stated in words as follows: A carry-out from a stage (which, of course, is equal to the carry-in of the next stage) will be produced if either one or both of the inputs are a logic '1' *or* if there was a carry-in to this stage and at least one of the inputs was a logic '1'.

Since the carry-out of a stage is the same as the carry-in of the next stage, we can write

$$CI_{i+1} = G_i + P_i.CI_i$$

This means that we can now write out the expression for the carry-in of any of the stages by starting with the first stage and expanding the expression for CI. For example,

Stage 0
$$CI_1 = G_0 + P_0.CI_0$$
$$= X_0.Y_0 \qquad \text{(assuming } CI_0 \text{ is '0')}$$

Stage 1
$$CI_2 = G_1 + P_1.CI_1$$
$$= X_1.Y_1 + (X_1 + Y_1).CI_1$$
$$= X_1.Y_1 + (X_1 + Y_1).(X_0.Y_0)$$

Stage 2
$$CI_3 = G_2 + P_2.CI_2$$
$$= X_2.Y_2 + (X_2 + Y_2).CI_2$$
$$= X_2.Y_2 + (X_2 + Y_2).(X_1.Y_1 + (X_1 + Y_1).(X_0.Y_0))$$

This process can be repeated for additional stages of a multi-bit adder.

The resulting expressions can be rewritten as two-level sum-of-products expressions. For example, the carry-in for stage 2 is equal to

$$CI_2 = X_1.Y_1 + X_1.X_0.Y_0 + X_0.Y_1.Y_0$$

which could be implemented with the logic circuit shown in Figure 11-2.

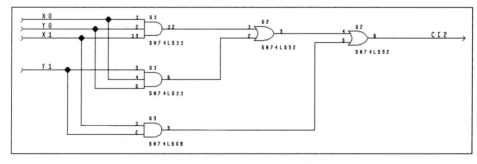

Figure 11-2: *Circuit diagram showing CI$_2$ generation*

The logical expression for the sum bits for each stage is simply

$$\text{SUM}_i = (X_i \oplus Y_i) \oplus CI_i$$

Using these expressions we can now design our look-ahead carry adder. The maximum delay for this type of adder should be equal to three logic gates, irrespective of the word length of the adder. This is a big improvement over the ripple-carry adder, where the delay increases as we increase the number of stages in the adder. However, it should be noted that when you convert the logic expressions of your carry adder into gates contained in 74LS devices you will find that you will need devices like an OR-gate with 6 or more inputs. Unfortunately, these devices do not exist, so you will need additional gates to implement the equivalent functionality, which will incur further delays. This would not be a problem for an integrated circuit manufacturer making this type of circuit, since the required gates could be made directly with the necessary inputs.

Although we have designed a circuit whose delays are independent of the number of stages, it should be obvious that the disadvantage of the look-ahead carry technique is the increased complexity of logic required to implement it. This is typical of the kind of trade-offs one has to make when designing digital and computer systems.

Procedure

Design a four-bit look-ahead carry adder with two four-bit inputs, X and Y, and a four-bit SUM output.

Verify that your circuit works correctly by using the *ViewSim* and *ViewTrace* tools to simulate the addition of the values shown in Table 11-2.

X[3:0]		Y[3:0]		SUM[3:0]	
Decimal	Binary	Decimal	Binary	Decimal	Binary
2	0010	3	0011		
5	0101	1	0001		
-4	1100	6	0110		
-1	1111	-5	1011		
1	0001	-1	1111		

Table 11-2: Verifying correct operation of a four-bit look-ahead carry adder

Using the last addition operation from the table (1 add -1), obtain a timing diagram showing the delay from applying the inputs to each of the SUM outputs (0, 1, 2, and 3) becoming stable. Start by applying 0000 and 1111 on the X and Y buses, respectively, simulate until the circuit is stable, then change the value on the X bus to 0001. This sequence is the same as that applied to the ripple-carry adder from Exercise 2. Comment on how these delays compare with the similar timing analysis you carried out using the ripple-carry adder.

The following figure shows the contents of a command file (see Chapter 6 for more details on command files) that you could use to simulate your design. Notice the pattern commands to apply the different values to the X and Y buses in sequence. The last two sets of patterns applied are to obtain the timing delay from changing the X bus from 0000 to 0001.

```
| Typical delays in use.
| All delays scaled by 1.
| Total of 20 digital modules were processed.
wave LOOKAHD.wfm x0 x1 x2 x3 y0 y1 y2 y3 sum0 sum1 sum2 sum3
vector x x[3:0]
vector y y[3:0]
vector sum sum[3:0]
watch x y sum
pattern x 0 0010 0101 1100 1111 0001 0000 0001
pattern y 0 0011 0001 0110 1011 1111 1111 1111
run
```

Figure 11-3: Example of command file (lookahd.cmd) for the look-ahead carry circuit

12 Laboratory Exercise 5: Design of a Four-Bit ALU

Description

The goal of this exercise is to extend the four-bit arithmetic unit from Laboratory Exercise 3 to include the capability to carry out Boolean operations on the two inputs. We will provide the capability to be able to carry out a logical AND or a logical OR operation on the individual bits of the X and Y inputs. The resulting design will be a simplified example of an Arithmetic and Logic Unit (ALU) which is capable of carrying out two arithmetic operations (addition and subtraction), and two logic operations (AND and OR). The selection of one of these four operations will be determined by the code applied to two control signals.

Theoretical Background

ALUs are available as individual commercial MSI devices (for example, the 74LS181, 74LS381, and 74LS382 4-bit ALUs), or they may be part of the logic of a microprocessor or microcontroller. The logic symbol of a 74LS381 Arithmetic Logic Unit/Function Generator is shown in Figure 12-1.

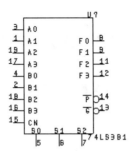

Figure 12-1: Logic symbol for 74LS381

There are two 4-bit operands (labeled A0-A3 and B0-B3) and a carry-in (CN) signal. The arithmetic or logic operation to be performed is selected by applying a code to the three select lines, S0, S1, and S2. The resulting output is available on the four output lines, labeled F0, F1, F2 and F3. The 74LS381 provides two carry lookahead signals, P' and G', for the four-bit ALU. If operations larger than four bits are required, multiple ALUs may be cascaded together with a look-ahead carry generator (for example, the '182), which will take the P' and G' signals from all the stages and produce the required carry-in signals.

The eight binary arithmetic and logic functions that may be performed by the '381 and the '382 are shown in Table 12-1. (The only difference between the '381 and the '382 is that the '381 produces group-carry look-ahead outputs, as mentioned above, while the '382 provides ripple carry and overflow outputs.)

Selection			Arithmetic/Logic
S2	S1	S0	Operation
0	0	0	CLEAR (F = 0000)
0	0	1	B MINUS A
0	1	0	A MINUS B
0	1	1	A PLUS B
1	0	0	A ⊕ B
1	0	1	A + B
1	1	0	A.B
1	1	1	PRESET (F = 1111)

Table 12-1: Function table of '381 and '382 4-bit ALUs

As can be seen from Table 12-1, the code applied to the three select inputs of the ALU determines the operation to be performed. In your ALU design, you have only two select inputs to determine which one of the four operations will be performed. These select lines can be decoded by a '138 or '139 type decoder to provide one active signal that may be used to select the circuitry required for the operation.

Procedure

The 4-bit ALU that you design in this laboratory exercise is similar to the operation of the '381 and '382. Modify your arithmetic unit design from Laboratory Exercise 3 to include two selection signals, S0 and S1, which will be used to determine the operation to be performed by your ALU as described in Table 12-2.

Selection		Arithmetic/Logic
S1	S0	Operation
0	0	Addition (X PLUS Y)
0	1	Subtraction (X MINUS Y)
1	0	AND (X.Y)
1	1	OR (X+Y)

Table 12-2: Function table of ALU

You will have to design and implement additional circuitry to perform the AND and OR operations along with some type of decoder and selection circuitry to select the operation required.

Demonstrate the correct operation of your circuit by using *ViewSim* and complete the following table.

Operation	SUM[3:0]
3 add 2	
4 subtract -3	
1011 AND 0010	
1110 AND 0001	
1010 OR 0101	
0001 OR 1000	

Table 12-3: Operation of ALU

CHAPTER

13 Laboratory Exercise 6: Design of a Random Number Generator

Description

The design exercises you have completed so far have all been examples of *combinational* circuits, circuits whose outputs are dependent only on the inputs currently applied. We now want to investigate the design of *sequential* circuits, which are circuits that contain storage elements and whose outputs depend not only on the inputs being currently applied but also on the past condition or *state* of the circuit. Such circuits are said to exhibit memory in that they remember their previous state and base their next actions on this information.

Although there are many different types of sequential circuits, there is a class of sequential circuits known as finite state machines (FSMs) that are relatively easy to understand and design. A block diagram of a finite state machine identifying the three main components is shown in Figure 13-1.

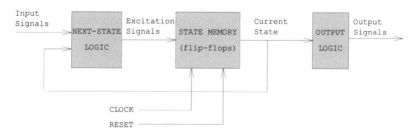

Figure 13-1: Block diagram of a finite state machine

The next-state logic and the output logic sections are combinational circuits, while the state memory section consists of a number of storage devices, usually flip-flops, that store the current state of the state machine. On each active clock transition (for example, the positive clock edge) the state machine will move from one state to another state based on the current inputs.

This laboratory exercise introduces you to the design of state machines by requiring you to design a random number generator that will simulate the operation of a die (i.e., a cube with the faces bearing 1 to 6 spots used for playing board games and in gambling. Note that the term dice is the plural of die but is sometimes used to refer to a single, as well as multiple, die.) The idea is that the circuit should display at high speed the digits 1 through 6 on a series of LEDs (light emitting diodes) when a 'roll' switch is pressed. On releasing the switch, the display should stop and display one of the digits. This circuit would thus be equivalent to rolling a die, which has an outcome of 1 to 6, depending on which side the die lands. Each outcome (1, 2, 3, 4, 5, or 6) should be equally likely. If not, you have designed the equivalent of a loaded die!

The circuit will have six outputs to control six LEDs and one input connected to the 'roll' switch. A block diagram of the system is shown in Figure 13-2.

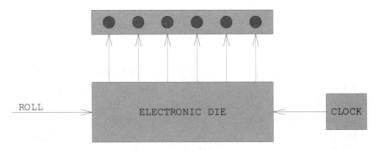

Figure 13-2: Block diagram of a random number generator (electronic die)

Theoretical Background

The simplest way to implement this design is with a clocked synchronous state machine arranged as a counter counting from 1 to 6. A two-bit counter can have a maximum of four states and hence can count to four, while a three-bit counter can have a maximum of eight states and can count to a maximum of eight. Your design should consist of three flip-flops for the state memory and simple combinational logic for the next-state logic and the output-logic sections.

We can produce a state/output table showing the next state and the output values for each of the possible current states, as shown in Table 13-1. Notice that when the Roll input is a logic '0', the count is held, and when Roll is a logic '1', the state machine increments to the next state, which increments the count.

Current	Next State		LED	LED	LED	LED	LED	LED
State	Roll = 0	Roll = 1	1	2	3	4	5	6
S0	S0	S1	1	0	0	0	0	0
S1	S1	S2	0	1	0	0	0	0
S2	S2	S3	0	0	1	0	0	0
S3	S3	S4	0	0	0	1	0	0
S4	S4	S5	0	0	0	0	1	0
S5	S5	S0	0	0	0	0	0	1

Table 13-1: *State/output table for a random number generator*

This design has six states, so we need three flip-flops resulting in two unused states (2^3 = 8 possible states with 3 flip-flops). We can assign the required states to the binary state variables represented by the flip-flops ($Q2$, $Q1$, and $Q0$) in a variety of different ways. For simplicity we will just assign the states in a binary order as shown in the following state assignment table.

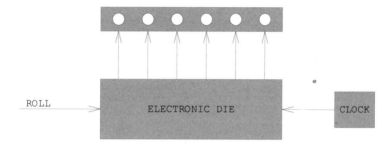

Table 13-2: *State assignments for a random number generator*

Once we have decided on a state assignment we can produce an excitation table based on the type of flip-flops we are going to use to implement our state machine. It is recommended that you use only D-type flip-flops, since the excitation equations are simply $Q^* = D$, which means that the next state of the flip-flip after the next active clock transition will be just equal to the value applied to the D input. The excitation table for the random number generator using D flip-flops is:

Current State	Next State (D2 D1 D0)	
(Q2 Q1 Q0)	Roll = 0	Roll = 1
000	000	001
001	001	010
010	010	011
011	011	100
100	100	101
101	101	000

Table 13-3: *Excitation table for a random number generator*

Once the excitation table has been obtained it is relatively straightforward to determine the excitation equations. Taking each flip-flop in turn, the excitation table is used to produce a truth table for each variable. For example, taking D0:

Q2 Q1 Q0 Roll	D0
0 0 0 0	0
0 0 1 0	1
0 1 0 0	0
0 1 1 0	1
1 0 0 0	0
1 0 1 0	1
0 0 0 1	1
0 0 1 1	0
0 1 0 1	1
0 1 1 1	0
1 0 0 1	1
1 0 1 1	0

Table 13-4: Truth table for D0

From this truth table it is possible to determine the next-state logic for D0 in a variety of ways. We will use a Karnaugh map to find the minimal sum-of-products expression for D0.

Q2 Q1 \ Q0 ROLL	00	01	11	10
0 0	0	1	0	1
0 1	0	1	0	1
1 1	D	D	D	D
1 0	0	1	0	1

Table 13-5: Karnaugh map for D0, with unused states shown as D, "don't care"

From this map we can determine the excitation equation for D0, which is simply

$$D0 = Q0'.ROLL + Q0.ROLL'$$

The same process can be used to determine excitation equations for D1 and D2.

Once the next-state logic has been completed, we can determine the output logic required to turn on and off the six LED's during the appropriate states. The state/output table (Table 13-1) can be rewritten in terms of the state assignments of Table 13-2 to obtain truth tables for each of the outputs. For example, the table for LED1 is

Current State Q2 Q1 Q0	LED 1
0 0 0	1
0 0 1	0
0 1 0	0
0 1 1	0
1 0 0	0
1 0 1	0

Table 13-6: Truth table for output LED1

From this table it is straightforward to obtain the output equations. Each output equation is simply the logic expression of one of the minterms. There is no minimization required for these output signals, so using Karnaugh maps or other techniques are unnecessary. For example, the expression for LED1 is simply

```
LED1 = Q2' Q1' Q0'
```

The equations for the other outputs may be obtained in a similar manner.

Once the next-state and output logic have been determined a circuit may be designed to implement the required functionality. To help you get started with your design, Figure 13-3 shows the next-state logic for flip-flop Q0 and the output logic for LED1. Notice the active-low Preset and Clear signals indicated by the horizontal bar across the label. Refer to Chapter 6 if you do not remember how to change the sense of a signal label.

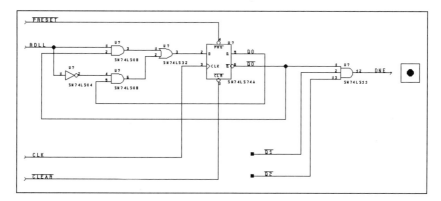

Figure 13-3: Circuit diagram showing flip-flop 0, with corresponding excitation signal and output logic

> **NOTE:** Although we have called this laboratory exercise the design of a random number generator, we should mention that what you have really designed is a three-bit synchronous binary counter that counts from 0 to 5 (when the Roll input is active). However, if the clock is sufficiently high, then the circuit can be used as an approximation to a random number generator.

Procedure

Design a state machine to implement the random number generator described above. Use three D-type flip-flops for the state memory.

Add a clock input to the design which is connected to the CLK inputs of the three flip-flops. We need a sufficiently high clock so that the changes in outputs are fast enough. A 1-kHz clock should be sufficient.

Also connect the preset and clear inputs on the D-type flip-flops to a 'preset' and a 'clear' input, respectively. We will use these during simulation to ensure that our state machine is initialized correctly.

Simulate your design using *ViewSim* and *ViewTrace*. Use a clock period of 1000 ns during simulation. To ensure that your circuit is initialized correctly before you start clocking the flip-flops, issue the following *ViewSim* commands (don't type the comments on the right!):

```
H ~PRESET          | Set the preset input high
L ~CLEAR           | Set the clear input low
L ROLL             | Set the roll input low - disable roll
SIM 100NS          | Simulate for 100 ns (reset flip-flops)
H ~CLEAR           | Set the clear input high
H ROLL             | Set the roll input high - enable roll
SIM 100NS          | Simulate for 100 ns
CLOCK CLK 0 1      | Create a clock signal using
                   | the clk input
STEPSIZE 500NS     | Set clock period to be 1000 ns
CYCLE              | Simulate for one clock cycle
CYCLE 6            | Cycle for six more clock cycles
```

The first few commands are used to reset the flip-flops by forcing the CLR inputs on the 74LS74A low for a period of 100 ns.

Include *ViewTrace* output waveforms to show that your random number generator cycles correctly through the six possible states.

By forcing the Roll signal low (L ROLL) you can demonstrate that your circuit holds the count.

14

Laboratory Exercise 7: Design of a Simple Traffic Controller

Description

You are to design a traffic light controller to control an intersection. The intersection consists of two streets, one running north-south (called NS) and the other running east-west (called EW).

The circuit has to control just one of the traffic lights on the NS street and one traffic light on the EW street, as shown in Figure 14-1. Each traffic signal consists of a red, yellow, and green light. Each signal will cycle through red, green, yellow, and back to red. When one signal is green, the other will be red; when yellow, the other will be red; and when red, the other will be green. To simplify the design, and also to shorten the experimentation time, we will set the green, yellow, and red time periods to 4, 1, and 5 seconds, respectively.

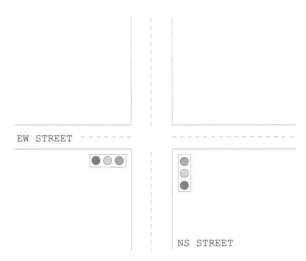

Figure 14-1: Diagram showing an intersection controlled by a traffic controller

Assume that you have a clock input that has a frequency of 1 Hz (1 cycle per second). You are to provide three outputs to control the traffic light on the NS street and three outputs to control the EW light.

Once you have designed the traffic controller you can demonstrate that it works correctly through simulation. Use *ViewSim* and *ViewTrace* to obtain a waveform showing the correct cycling of your traffic lights.

> **NOTE:** During simulation, instead of working with an input clock frequency of 1 Hz, use 1 MHz (1000-ns period).

Theoretical Background

It is recommended that you design your traffic controller as a clocked synchronous state machine with 10 states, and use one state for each of the 10 seconds required in a traffic light cycle. Refer to the previous chapter for information on designing state machines. In this design we do not have any inputs, apart from the clock and a reset line, to initialize the circuit. A block diagram of this state machine is shown below:

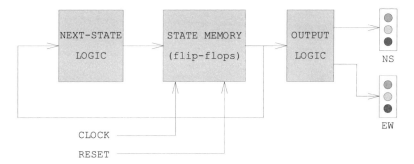

Figure 14-2: *Block diagram of a traffic controller state machine*

The first step in this design is to construct a state/output table that lists each of the possible current states along with the outputs and the next states. You may find it helps to initially draw a state diagram.

The number of flip-flops needed to code the 10 states is the smallest integer greater than or equal to $\log_2 10$, which is 4.

It is recommended that you use D-type flip-flops (for example, a 74LS74 contains two D-type flip-flops), since this will make your next state logic easier.

You should construct an excitation table that shows the input values needed for each of the four D-type flip-flops to move to the desired next state. From this table you can derive the excitation equations and simplify them by using Karnaugh maps.

From the state/output table you can derive the output equations to drive the six lights. These equations may also be simplified by using Karnaugh maps.

HINT: Careful selection of the 10 required states to your state variables may help to minimize the next-state and output logic required.

Procedure

Design a simple traffic controller with the specifications given above.

In your lab report you must include a section providing a functional description and explanation of your circuit. This should include details on the type of finite state machine you have designed, Boolean expressions with any resultant minimization, and other relevant information to explain your design.

Also required are schematics of your circuit and sufficient *ViewTrace* plots (or other type of documentation) to demonstrate the correct operation of your design. The waveform plots should show the condition of the three NS and EW lights as your state machine cycles through the 10 states.

15

Laboratory Exercise 8: Design of a Traffic Controller with Left Turn Signal

Description

This traffic controller design is very similar to the controller from Laboratory Exercise 7, but with the addition of a left turn traffic sensor and signal for the NS street as shown in Figure 15-1.

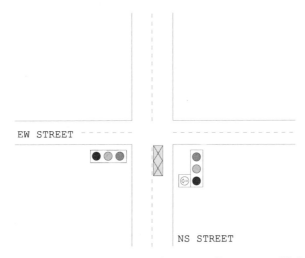

Figure 15-1: Intersection with a left turn traffic sensor and light

We assume that the sensor will produce a logic '1' signal anytime there is a car or cars waiting to turn left on the NS street. If no cars are detected then the traffic controller will simply cycle the NS and EW lights as in the previous design. At the end of the EW street's green and yellow period, when the NS street's green phase would normally start, the left turn detector is checked to see if a car is waiting to turn left. If a car is waiting then the left turn light should be turned on for two seconds, indicating that the car can now turn left. During these two seconds the EW and NS lights will still be at red. Once the two seconds have elapsed then the left turn light should be switched off and the NS green light set to on. The cycle then continues as before.

Theoretical Background

The design process for this controller is very similar to the previous laboratory exercise. The state machine will now have 12 states instead of 10 to support the two-second left turn condition. A table showing the 12 possible states is shown below. The letters G, Y, and R denote when the green, yellow, and red lights are on and the letter L denotes when the left turn indicator is on. The bottom line of the table represents the five distinct output states (OS) which have been labeled A, B, C, D, and E.

	1	2	3	4	5	6	7	8	9	10	11	12
NS	G	G	G	G	Y	R	R	R	R	R	L	L
EW	R	R	R	R	R	G	G	G	G	Y	R	R
OS			A		B		C			D	E	

Table 15-1: Traffic controller states

The previous design did not have any conditional inputs and so the excitation equations and resulting logic were simply a function of the current state. We now have an input representing the left turn car detector and if this input is active (indicating that a car is waiting) during state 10 then the state machine will go into state 11, then 12, then back to 1. If, on the other hand, the input is inactive (no cars waiting to turn left), then the state machine must go directly from state 10 to state 1.

Procedure

Design a traffic controller with the specifications given above.

In your lab report you must include a section providing a functional description and explanation of your circuit. This should include details on the type of finite state machine you have designed, Boolean expressions with any resultant minimization, and other relevant information to explain your design.

Also required are schematics of your circuit and sufficient *ViewTrace* plots (or other type of documentation) to demonstrate the correct operation of your design. The waveform plots should show the condition of the three NS and EW lights and the left turn light for a condition when no car is waiting to turn left and also when a car is waiting to turn left.

16 Laboratory Exercise 9: Design of a Sequential Combination Lock

Description

This laboratory exercise involves the design of a sequential combination lock similar in concept to security locks found on door entrances or safes. A sequence of numbers has to be entered in the correct order for the lock to be released. A mechanical version of a sequential combination lock requires a dial to be rotated clockwise or counterclockwise until the first number is aligned with a mark. Subsequent numbers are then entered by rotating the dial in alternate directions and aligning the new numbers. Once the correct combination of numbers has been entered, the lock mechanism releases and allows the device to which it is attached to be opened. An electronic equivalent usually has a numeric keypad through which the numbers must be entered one by one to open the lock.

We will design a simple electronic sequential lock. To reduce the size of the design we will require only four numbers to be entered and each of these numbers will be limited to the digits 0, 1, 2, or 3. We will assume that the lock will open when we enter the following sequence of numbers—2, 1, 3, 1.

We will apply the numbers by turning on one of four switches representing the four digits and then press a further switch (which we will use as a clock) to enter the new number. A single output signal will show the condition of the lock: A logic '1' will indicate the locked state, and a logic '0' will indicate the unlocked state.

Theoretical Background

It should be obvious that we need a sequential circuit to implement such a design. The system has to be able to remember the previous numbers entered, hence it requires some type of storage or memory capability. A synchronous state machine similar to the designs in the previous lab exercises can be used.

A state diagram is shown in Figure 16-1. It can be seen that there are four states, labeled S0, S1, S2, and S3. The input and resulting output are shown for each of the possible transitions from state to state. The notation is X/Y, where X indicates the input(s) and Y represents the output. It can be seen that the state machine advances to the next state only when the correct input number sequence is entered.

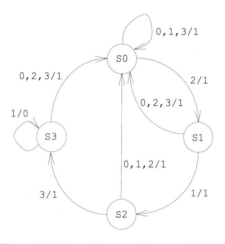

Figure 16-1: *State diagram of a sequential lock*

The same information can be displayed in the form of a state/output table, as shown in Table 16-1. The next state and lock signal values are shown for each of the current states and four possible input numbers.

	Next State and Lock Signal			
	Input Number			
State	0	1	2	3
S0	S0,1	S0,1	S1,1	S0,1
S1	S0,1	S2,1	S0,1	S0,1
S2	S0,1	S0,1	S0,1	S3,1
S3	S0,1	S3,0	S0,1	S0,1

Table 16-1: *State/output table for a sequential lock*

A block diagram of the state machine is shown in Figure 16-2.

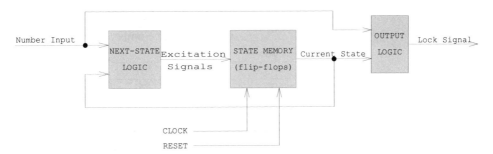

Figure 16-2: *Block diagram of a sequential lock state machine*

This type of state machine is known as a Mealy machine; the output is dependent on both the current state and the current input. In the sequential circuits designed in the previous laboratory exercises (random number generator and traffic controllers) the outputs were dependent only on the current state; these types of sequential circuits are called Moore machines.

Looking at the block diagram we see that we need to design the next-state logic, the state memory, and the output logic.

The usual place to start the design of a finite state machine is with the state memory. There are four possible states (S0, S1, S2, and S3) and so we need two flip-flops to code them. We can assign the states in a number of different ways but we use a simple binary sequence for this design, as shown in Table 16-2.

State Name	Q1 Q0
S0	0 0
S1	0 1
S2	1 0
S3	1 1

Table 16-2: *State assignment for a sequential lock*

As in the previous state-machine designs you have carried out, it is recommended that you use D-type flip-flops (because it makes the excitation logic simpler!) for the state memory.

Once the state-memory requirements have been determined we can start looking at the excitation equations required. The next-state is determined by the current state and also the current input conditions. The input number is entered by turning on one out of four switches. The active switch number can be encoded into a two-bit binary value. We thus have four variables—two for the current state and two for the input value.

The excitation table for the sequential logic is shown in Table 16-3. It can be seen that the data inputs to the two flip-flops are dependent on the current state and the current input.

State	D1 and D0			
	Encoded Input Number			
Q1 Q0	00	01	10	11
0 0	00	00	01	00
0 1	00	10	00	00
1 0	00	00	00	11
1 1	00	11	00	00

Table 16-3: Excitation table for a sequential lock

From the excitation table it should be possible to readily determine the excitation logic required. The minimum logic required may be obtained with the aid of Karnaugh maps or other logic minimization techniques.

The output logic can be obtained from Table 16-1 by replacing the state names with the state assignments made earlier.

Procedure

Design a sequential combination lock with the features described above. Assume that the number being entered has been encoded into a two-bit binary number (00, 01, 10, and 11) and is input on two lines, SW1 and SW0.

Demonstrate that your circuit works correctly by using the *ViewSim* and *ViewTrace* tools.

Apply various input combinations and show the sequential lock produces the correct output (lock or unlocked state) when you enter different number sequences.

The following is a command file that you may wish to use to test your design (refer to Chapter 6 to review the use of command files):

```
| Command file to test Sequential Lock
|
restart
|
| Reset flip-flops
h ~pre                        | set the preset input high
l ~clr                        | set the clear input low
sim 100ns                     | simulate for 100ns
h ~clr                        | set the clear input high
sim 100ns                     | simulate for a further 100ns
|
| combine sw0 and sw1 into a vector to make it easier to
| apply inputs
vector sw sw1 sw0
|
| make a clock with stepsize 100ns (period 200ns)
clock clk 0 1
stepsize 100ns
```

```
|
| now enter a sequence - last four should release lock
pattern sw 0\d 3\d 1\d 2\d 1\d 3\d 1\d
run
|
| now enter a new sequence - lock should be reactivated
pattern sw 1\d 3\d 2\d
run
```

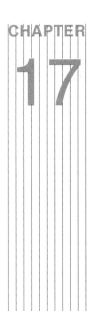

CHAPTER 17

Laboratory Exercise 10: Design of a Vending Machine

Description

This laboratory exercise involves the design of a simple vending machine that dispenses hot drinks. To reduce the size of the design we will restrict the number of choices that can be made and also ignore some operations. For example, we will ignore the circuitry required to accept coins and check when the required amount has been entered. However, once the basic design is complete it should be fairly easy to add more functionality.

The only drinks available in our simple vending machine will be coffee or tea, with or without whitener. The selection will be made on three switches. The outputs of the vending machine will be one light and five solenoids to control the dispensing of various items and to open the door. The inputs and outputs of the vending machine are shown in the block diagram in Figure 17-1.

Figure 17-1: Block diagram of a vending machine

The vending machine will have four possible states. The first state will turn on the Make Selection light to indicate that the required drink selection may be pressed. The vending machine will stay in this state until a selection is made. We will assume that the selection is latched into a suitable register and so once a selection has been pressed that input will stay active until the current dispensing cycle is completed.

Once either the coffee or tea selection is pressed the vending machine should move to the second state. In this state the Make Selection light should be turned off and the machine should start dispensing either coffee or tea as selected. After this state the machine should move to the third state and dispense hot water and also the whitener if this was selected previously. In the fourth state the machine should open the door to allow the drink to be removed. After this state the machine should go back to the first state for the next dispensing cycle.

All of the states should be equal in duration (we will assume they are one second in length) apart from the first state. The machine will stay in the Make Selection state until either a coffee or tea selection has been made.

Theoretical Background

A synchronous state machine similar to the designs in the previous lab exercises can be used to implement this design. The four states of the machine can be called S0, S1, S2, and S3.

A state/output table should be created similar to that shown in Table 17-1.

	Next State, Make Selection, Dispense Coffee, Tea, Whitener, Water, OpenDoor							
	Selection (coffee, tea, whitener)							
State	000	001	010	011	100	101	110	111
S0	S0,1,0,0, 0,0,0	S0,1,0,0, 0,0,0	S1,1,0,0, 0,0,0					
S1								
S2								
S3								

Table 17-1: State/output table for a vending machine

We will require a Mealy state machine, since the outputs are dependent on both the current state and the current inputs. A block diagram of the state machine is shown in Figure 17-2.

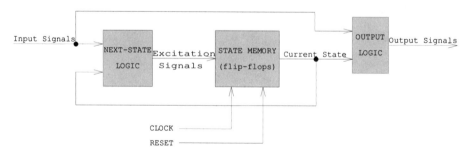

Figure 17-2: Block diagram of a vending machine

Looking at the block diagram we see that we need to design the next-state logic, the state memory, and the output logic.

The usual place to start the design of a finite state machine is with the state memory. There are four possible states (S0, S1, S2 and S3) and so we need two flip-flops to code them. We can assign the states in a number of different ways but we use a simple binary sequence for this design, as shown in the following table:

State Name	Q1 Q0
S0	0 0
S1	0 1
S2	1 0
S3	1 1

Table 17-2: *State assignment for a vending machine*

As in the previous state machine designs you have carried out, it is recommended that you use D-type flip-flops (because it makes the excitation logic simpler!) for the state memory.

Once the state-memory requirements have been determined we can start looking at the excitation equations required. The next state is determined by the current state and also the current input conditions. There are three possible inputs for the selection of coffee, tea, and whitener. Remember that we can assume that these inputs are latched, which means that once a selection has been made, the corresponding input will remain active during the remainder of the vending cycle. Using these inputs to determine that a selection has been made allows us to move from state S0 to S1. The same inputs will also tell us which selections are made during states S2 and S3, so that the correct outputs will be activated.

Once an excitation table has been developed, it should be possible to readily determine the excitation logic required. The minimum logic required may be obtained with the aid of Karnaugh maps or another logic minimization technique. Finally, the output logic can be obtained from the same table.

For more information on the design of a Mealy state machine, review the design of the sequential lock in the previous laboratory exercise.

Procedure

Design a vending machine with the features described above. Assume the clock input has a period of 1 second. Demonstrate your circuit works correctly by using the *ViewSim* and *ViewTrace* tools.

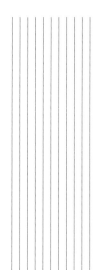

Index of Commands

Menu Commands